# Advances in Geophysical and Environmental Mechanics and Mathematics

Series Editor: Professor Kolumban Hutter

Karl-Heinz Glaßmeier · Heinrich Soffel ·
Jörg F.W. Negendank

# Geomagnetic Field
# Variations

 Springer

Karl-Heinz Glaßmeier
TU Braunschweig
Institut für Geophysik und
extraterrestrische Physik
Mendelssohnstr. 3
38106 Braunschweig
Germany
kh.glassmeier@tu-bs.de

Heinrich Soffel
Department für Geo- und
Umweltwissenschaften
Universität München
Theresienstr. 41
80333 München
Germany
heinrich.soffel@geophysik.
uni-muenchen.de

Jörg F.W. Negendank
GeoForschungsZentrum Potsdam
Telegrafenberg
14473 Potsdam
Germany
neg@gfz-potsdam.de

*Cover inlay image: Intensity of the secular variation 2003.0 of the geomagnetic field.
Courtesy: GeoForschungsZentrum Potsdam*

ISBN: 978-3-642-09560-3          e-ISBN: 978-3-540-76939-2

Advances in Geophysical and Environmental Mechanics and Mathematics

ISSN: 1866-8348     e-ISSN: 1866-8356

*Cover design:* deblik, Berlin

Printed on acid-free paper

9 8 7 6 5 4 3 2 1

springer.com

*To all those colleagues operating the worldwide network of geomagnetic observatories. Without their careful work, scientific studies on the geomagnetic field would be impossible.*

# Preface

Earth's magnetic field is currently changing dramatically. The dipole moment decreased by about 10% since the times of Carl-Friedrich Gauss. The most drastic change of the geomagnetic field is a polarity transition – an event during which the magnetic poles reverse their signs. During such a transition the field magnitude at the Earths surface diminishes to about 10% of its normal value. The last reversal occurred about 780,000 years ago. Is the observed decrease of the dipole moment indicating a future polarity transition? What would be the effects of such a drastic change to system Earth? Can any positive or negative effects to our biosphere or even humans be expected?

Between 2000 and 2006 the Deutsche Forschungsgemeinschaft (DFG) conducted a special priority programme *Geomagnetic Variations: Space-Time Structure, Processes, and Effects on System Earth* to unravel some of the open questions related to geomagnetic field changes. The launch of the satellite CHAMP in the year 2000 with its magnetometer systems on board and the much improved ability to analyse the global magnetic and its temporal variation was one of the reasons to initiate this DFG research programme.

Almost all research groups in Germany with interest and experience in problems of geomagnetism and related fields were involved. The programme was coordinated by the editors of this book and guided by a board of referees of the DFG consisting of a group of highly esteemed scientists from various countries, who accompanied the project from the beginning in 2000 to its final symposium in October 2006 in Braunschweig.

The overall programme was very successful in bringing together a large number of scientists from very different disciplines such as geomagnetism, paleomagnetism, geology, theoretical physics, solar physics, astrophysics, applied mathematics, magnetospheric physics, and atmospheric physics. The results of the programme have been published in some 150 refereed publications and numerous oral presentations.

We are very grateful to the Deutsche Forschungsgemeinschaft for the financial support of this research programme and, in particular, to Dr. Johannes Karte, for his professional guidance. Special thanks are to our national and international colleagues in the group of referees, Eigil Friis-Christensen, Rainer Hollerbach, Gauthier Hulot, Andy Jackson, Dominique Jault, Cor Langereis, Bob McPherron,

John Shaw, Jürgen Untiedt, Kathy Whaler, and Helmut Wilhelm for their continuous support and constructive advice.

The intention of this book is not to be a textbook in geomagnetism, but to review major results of the above mentioned programme and other ongoing international activities aiming at understanding the causes and effects of geomagnetic variations. As the topic of the book covers a wide range of scientific disciplines, the first chapter aims at summarizing basic principles of geomagnetism and related fields in order to ease the reading of the later, more specialized chapters reviewing current scientific knowledge. These later chapters have been written by key people in the respective research areas, supported by a large number of other experts in the field. In particular contributions by Valerian Bachtadse, Ulrich Bleil, Jan-Philipp Bornebusch, John P. Burrows, Martyn Chipperfield, Charles H. Jackman, Kumar Hemant, Anne Hemshorn, Christoph Heunemann, Daniela I. Hofmann, David Krása, Klaus F. Künzi, Stefan Maus, Justus Notholt, Norbert Nowaczyk, Nicolai Petersen, Jens Poppenburg, Patricia Ritter, Martin Rother, Jens Schroeter, Anja Stadelmann, Friedhelm Steinhilber, Claudia Stolle, Christian Vérard, Holger Winkler, Michael Winklhofer, Ingo Wardinski, Jan Mark Wissing, and Bertalan Zieger are acknowledged.

Braunschweig                                                             *Karl-Heinz Glaßmeier*
June, 2008                                                                     *Heinrich Soffel*
                                                                              *Jörg Negendank*

# Contents

# Contents

# List of Contributors

Karl Fabian
Geological Survey of Norway Leiv Eiriksons vei 39 7491 Trondheim Norwegen

Karl-Heinz Glaßmeier
Institut für Geophysik und extraterrestrische Physik Technische Universität Braunschweig Mendelssohnstrasse 2 38106 Braunschweig Germany

Helmut Harder
Institut für Geophysik Universität Münster Corrensstrasse 24 48149 Münster Germany

May-Britt Kallenrode
Fachbereich Physik Universität Osnabrück Barbarastrasse 7 49076 Osnabrück Germany

Monika Korte
GeoForschungsZentrum Potsdam Telegrafenberg 14473 Potsdam Germany

Roman Leonhardt
Department Angewandte Geowissenschaften und Geophysik Lehrstuhl Technische Ökosystemanalyse Universität Leoben Peter Tunner Straße 25-27 8700 Leoben Austria

Hermann Lühr
GeoForschungsZentrum Potsdam Telegrafenberg 14473 Potsdam Germany

Mioara Mandea
GeoForschungsZentrum Potsdam Telegrafenberg 14473 Potsdam Germany

Jörg Negendank
GeoForschungsZentrum Potsdam Telegrafenberg 14473 Potsdam Germany


Miriam Sinnhuber
Institut für Umweltphysik Universität Bremen Otto-Hahn-Allee 1 28359 Bremen
Germany


Heinrich Soffel
Ludwig-Maximilians-Universität München Fakultät für Geowissenschaften
Department für Geo- und Umweltwissenschaften Theresienstraße 41 80333
München Germany


Stephan Stellmach
Earth Sciences Department University of California Santa Cruz, CA 95064 USA


Joachim Vogt
Jacobs University Bremen School of Engineering and Science Campus Ring 1
28759 Bremen Germany


Johannes Wicht
Max-Planck-Institut für Sonnensystemforschung Max-Planck-Straße 2 37191
Katlenburg-Lindau Germany

# Chapter 1
# The Geomagnetic Field

Karl-Heinz Glaßmeier, Heinrich Soffel and Jörg Negendank

## 1.1 Introduction and Historical Notes

Since more than 3.7 billion years, planet Earth has a significant global magnetic field. It is most probably generated in the outer fluid core of our planet, and it varies on many different time scales, millions of years to seconds. The long-term variations are due to changes in the dynamo region, while the shorter term variations have their origin in electric current systems in the upper atmosphere and magnetosphere. These geomagnetic variations are a fascinating scientific topic as many different physical processes are involved, some of them still not well understood. Much effort is put into the study of the space–time structure of these variations and their underlying physics. In recent years also etc, the question about possible impacts of geomagnetic variations on system Earth and even the biosphere gains much more interest. Some organisms, migrating birds and pigeons, are known to use the geomagnetic field for orientation (e.g. Wiltschko and Wiltschko, 2005) and also bats are reported to respond to magnetic field polarity (Wang et al., 2007). Geo-biomagnetism is an emerging field (e.g. Winklhofer, 2004).

Furthermore, hypothetical connections between climate change, geomagnetic variations, and the collapse of civilizations are discussed as well (e.g. Courtillot et al., 2007; Haug et al., 2003). Also, a predisposition of part of the human population to adverse health due to geomagnetic variations is discussed in the scientific literature (Palmer et al., 2006). As some of these hypotheses are speculative, they cause a renewed and ongoing interest in geomagnetic variations, their space-time structure, and possible effects on system Earth.

Karl-Heinz Glaßmeier
Institut für Geophysik und extraterrestrische Physik Technische Universität Braunschweig Mendelssohnstrasse 2 38106 Braunschweig Germany

Heinrich Soffel
Ludwig-Maximilians-Universität München Fakultät für Geowissenschaften Department für Geo- und Umweltwissenschaften Theresienstraße 41 80333 München Germany

Jörg Negendank
GeoForschungsZentrum Potsdam Telegrafenberg 14473 Potsdam Germany

K.-H. Glaßmeier et al. (eds.), *Geomagnetic Field Variations*, Advances in Geophysical and Environmental Mechanics and Mathematics,
© Springer-Verlag Berlin Heidelberg 2009

Geomagnetism is the oldest discipline in geophysics. Trustworthy reports from the Han-epoch in China, about 202–220 BC, tell us about the tschinan-tsche, the noon-pointing cart, that is a cart carrying a compass (Balmer, 1956). And ever since the compass has been used to determine the north direction using the magnetic field, the origin of the magnetic field has been a mystery and is even still a mystery today for most people. It was Pierre de Maricourt, called Peregrinus, who as early as 1269 postulated in his *Epistola Petri Peregrini de Maricourt ad Sygerum de Foucaucourt, Militem, de Magnete* (de Maricourt, 1902) that the geomagnetic field originates within the Earth's interior and is associated with its rotation. In the late 15th century, William Gilbert compiled the knowledge of his time about magnets and the geomagnetic field in his famous book entitled *De Magnete, magneticisque corporibus, et de magno magnete tellure; Physiologia nova, plurimis et argumentis demonstrata* (Gilbert, 1600), which was published in 1600. He also concluded that Earth itself is a huge magnet and that its magnetic field is generated within the Earth. And he coined the famous sentence *Magnus Magnes Ipse est Globus Terrestris* (Fig. 1.1) being convinced that the interior of the Earth consists mainly of iron.

The geomagnetic field is a vector field, which can be described on the Earth's surface by its three orthogonal components or magnetic elements $X$ (pointing in the geographic north direction), $Y$ (pointing eastward), and $Z$ (pointing downward). The two horizontal components $X$ and $Y$ can be combined, yielding the horizontal component $H$ with $H = \sqrt{X^2 + Y^2}$, which is aligned in the direction of the compass needle. By also adding the vertical component $Z$, the intensity of the total field $F$ is obtained as $F = \sqrt{X^2 + Y^2 + Z^2}$. The declination $D$ is defined as the angle between $H$ and geographic north and the inclination is the angle between the horizontal

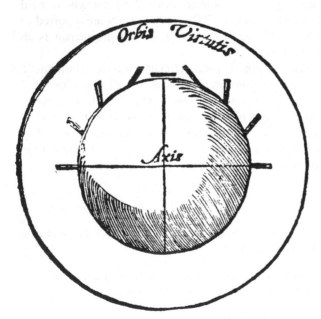

**Fig. 1.1** Spherical magnet with short magnets indicating the direction of the magnetic field at its surface (from Gilbert, 1600)

**Fig. 1.2** Geomagnetic coordinate system showing the three orthogonal components $X$, $Y$, and $Z$, the horizontal component $H$ and the total field $F$ as well as the declination D and the inclination I

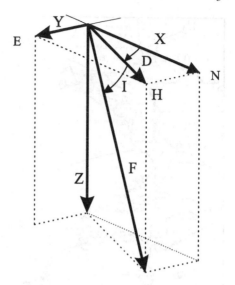

plane and the field vector $F$. For further illustration see Fig. 1.2. The magnetic field strength is nowadays measured in the unit Tesla (T), with $1T = 1\ Vs/m^2$. In geomagnetism subunits like $10^{-6}T = 1\ \mu T$ or $10^{-9}T = 1\ nT$ are more convenient, because the strength of the geomagnetic field in mid latitudes is about 50 $\mu T$ or 50,000 nT. Other units which have been used in the literature in the past are Gauss $(g^{1/2}cm^{-1/2}s^{-1})$ and gamma with $10^{-5}$Gauss = 1 $\gamma$. The relationships between the units are as follows: 1 Gauss = $10^{-4}$ T and 1 nT is $1\gamma$.

About hundred years after Gilbert, Edmond Halley published maps of the declination variations of the Atlantic region (1701) and of the entire world (1702) using data measured by himself with his research vessel *Paramore* in the Atlantic region, adding those compiled by the numerous ships crossing the oceans during the previous two centuries (Halley, 1701). Halley's maps (Fig. 1.3) stimulated other scientists afterward to produce maps containing other global geophysical data. One of them was, for instance, Alexander von Humboldt. Together with Carl Friedrich Gauss he founded the *Göttinger Magnetischer Verein* (Göttingen Magnetic Union) in 1836, aiming at a global network of observatories to study the geomagnetic field and its regional and temporal variations more in detail. This initiative resulted in the foundation of many new geomagnetic observatories in all parts of the world during the following decades. Carl-Friedrich Gauss and Wilhelm Weber used all available information of the geomagnetic field of their time and presented them in 1840 in their *Atlas des Erdmagnetismus* (Gauss et al., 1840), a collection of global maps like the one shown in Fig. 1.4 for the declination values.

It would lead too far in this book, aiming at the presentation of new results about the geomagnetic field, to discuss in more detail the early history of geomagnetism. For this we refer to monographs like those by Sydney Chapman and Julius Bartels (Chapman and Bartels, 1940), Heinz Balmer (Balmer, 1956), and Walter Kertz (Kertz, 1999), just to name a few. More historical notes can also be found in the various chapters of this book.

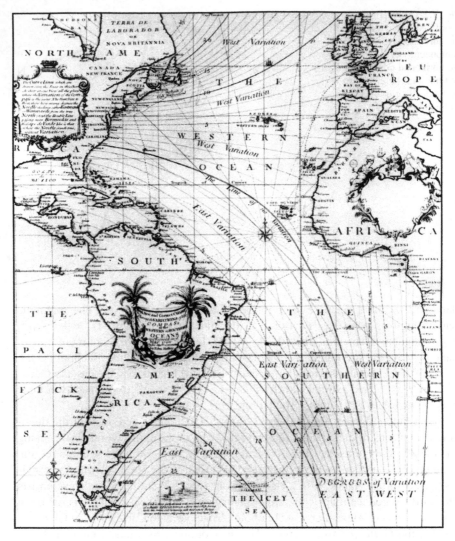

**Fig. 1.3** Map published by Edmond Halley in 1701 showing the lines of constant declination values (isogones) in the Atlantic ocean (from Halley, 1701)

## 1.2 The Recent Geomagnetic Field

The most simple approximation of the geomagnetic field is that of a hypothetic magnet (dipole) in the center of the Earth, with its axis anti-parallel to the axis of rotation (geocentric axial dipole). Such a model was already proposed by William Gilbert in 1600 (Gilbert, 1600). In this case the needle of a compass everywhere points northward and the declination is zero all over the globe. The inclination $I$ is $+90°$ at the geographic north pole, $-90°$ at the geographic south pole, respectively, and $0°$ at the

**Fig. 1.4** Global map of the declination, compiled and published by Gauss et al. (1840)

equator. At sites with the geographic latitude $\phi$, the relation between the inclination $I$ and latitude is given by $\tan I = 2\tan \phi$. At the equator the intensity of the field is only half of the intensity at the poles. However, the geocentric axial dipole is just a crude approximation. A better description of the present day geomagnetic field is that of a geocentric dipole, with the dipole axis tilted about $11°$ with respect to the axis of rotation (Fig. 1.5).

A more sophisticated method for the description of the geomagnetic field was introduced by Gauss in 1839 using spherical harmonic functions (Gauss, 1839).

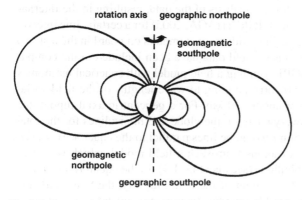

**Fig. 1.5** Geocentric dipole and its field lines. The dipole axis is tilted about $11°$ with respect to the axis of rotation. The physical definition of magnetic poles is used

Assuming that the magnetic vector field can be described as the spatial gradient of a scalar magnetic potential $V_i$, this potential can be represented as

$$V_i(r,\theta,\phi) = \sum_{n=1}^{\infty} \sum_{m=0}^{n} R_E \left(\frac{R_E}{r}\right)^{n+1} \left[g_{i;n}^m \cos m\phi + h_{i;n}^m \sin m\phi\right] P_n^m(\theta), \qquad (1.1)$$

where $P_n^m$ are the Schmidt quasi-normalized associated Legendre functions of degree $n$ and order $m$. The coefficients $g$ and $h$ are now called Gauss coefficients. Geographic longitude and latitude are denoted by $\phi$ and $\theta$, respectively; $R_E$ is the Earth's radius and $r$ the distance to the center of the Earth. The potential described in this way is the potential of magnetic sources located within the Earth interior.

At the Earth's surface fields originating outside, that is fields due to ionospheric and magnetospheric currents, can be represented by the potential $V_e$:

$$V_e(r,\theta,\phi) = \sum_{n=1}^{\infty} \sum_{m=0}^{n} R_E \left(\frac{r}{R_E}\right)^{n} \left[g_{e;n}^m \cos m\phi + h_{e;n}^m \sin m\phi\right] P_n^m(\theta). \qquad (1.2)$$

With this representation the field observed at the Earth's surface can be subdivided into fields originating from dipoles and combinations of dipoles (quadrupoles, octupoles, more generally called multipoles) centered inside or outside the Earth. Gauss was able to show that most of the field originates from within the Earth and that the dipole component yields the largest contribution (about 90%). Only a small contribution has its origin outside the Earth. However, in those days such external contributions could not yet be explained. Nowadays we know that electrical current systems in the ionosphere and magnetosphere are responsible for these magnetic fields.

Since the first analysis by Gauss the dipole moment of the geomagnetic field has lost about 10% of its intensity (Fig. 1.6) and the declination in Europe changed by about 20° (Fig. 1.7). Such effects are called secular variation and have already been noticed much earlier.

Using the data of the world wide network of roughly 100 observatories, the geomagnetic field has been analyzed regularly during the last 150 years and the data have been published as global maps by various authors. Since about 40 years standardized methods were applied for the analysis of the data, resulting in the International Geomagnetic Reference Field (IGRF) (IAGA, 2005) for a certain time interval (Fig. 1.8). After the advent of satellites with magnetometers on board in the last two decades of the 20th century, magnetic field data have been included in the computation of new versions of the IGRF, yielding a field model with spherical harmonics up to degree and order about 80 corresponding to wave lengths of the field variations and a spatial resolution of about 500 km. For a geocentric axial dipole, the declination $D$ would be zero everywhere on the globe and the isolines for the total intensity $F$ and for the inclination $I$ would be lines parallel to the equator. However, Figs. 1.8 and 1.9 show great deviations from such a field model, which is indicative for strong nondipole contributions (see also Fig. 1.4). In the equatorial southern Atlantic, there is a pronounced minimum of the field intensity, the South Atlantic Anomaly. In contrast to that, the field in the Pacific region is much less disturbed.

**Fig. 1.6** Decay of the dipole moment of the geomagnetic field during the last 200 years. Data compiled from McDonald and Gunst (1968) and the IGRF

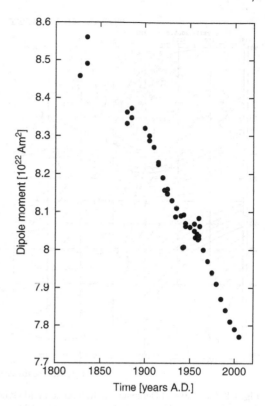

The German satellite CHAMP, orbiting the Earth at an altitude of about 450 km, was launched in the year 2000. It carries, among other instruments, two magnetometer systems. With its polar orbit a complete coverage of the Earth's surface is possible, yielding total intensity as well as vectorial data of the geomagnetic field. The results obtained from these data sets global data for the secular variation and secular acceleration for the time interval 2000–2007, new data for the electric current

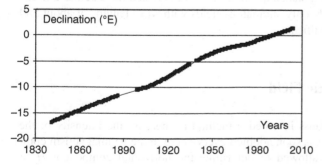

**Fig. 1.7** The geomagnetic declination measured at the Magnetic Observatory München-Maisach-Fürstenfeldbruck since 1840; it changed almost linearly from −17° to +2° between 1840 and 2007

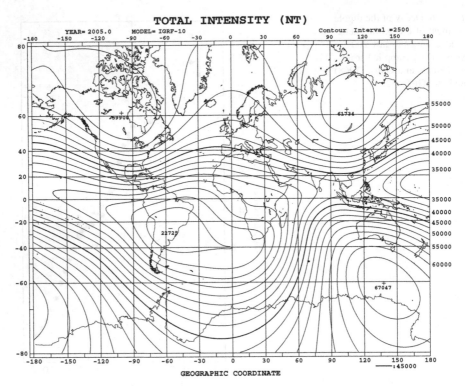

**Fig. 1.8** International Geomagnetic Reference Field (IGRF) for 2005, showing the global map of the total intensity. This map is by courtesy of the Data Analysis Center for Geomagnetism and Space Magnetism, Kyoto University, Japan. Further maps and representations can be found on the following webpage: http://swdcwww.kugi.kyoto-u.ac.jp/igrf/index.html

systems in the ionosphere, and even information about the contribution of oceanic currents are available and described in more detail in Chap. 2 of this book. For the first time it was also possible to produce a global map of the geomagnetic field contributions stemming from the lithosphere, mainly from the uppermost 20 km of the crust. However, due to the distance of about 450 km between the satellite altitude and the surface of the Earth, only anomalous fields with wave lengths of about 500 km and more could be identified.

## 1.3 The Paleomagnetic Field

In 1832, Carl-Friedrich Gauss invented a method to measure the intensity of the geomagnetic field, which was soon adopted everywhere in the geomagnetism community (Gauss, 1833). It allowed to determine the horizontal component $H$ of the field intensity at any desired point. So, together with the declination and the inclination, the entire vector **F** could be calculated. For times before 1832, no such

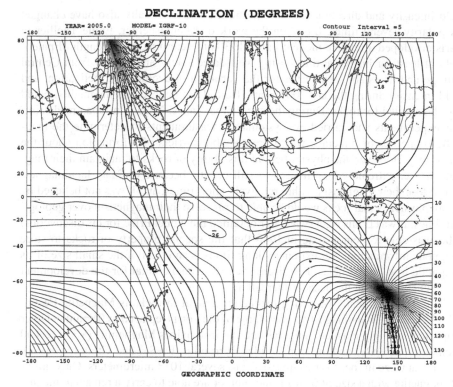

**Fig. 1.9** International Geomagnetic Reference Field (IGRF) for 2005, showing the global map of the declination. This map is by courtesy of the Data Analysis Center for Geomagnetism and Space Magnetism, Kyoto University, Japan

intensity values were available for a long time. Only few data could be obtained from the rarely made relative intensity measurements using the oscillation period of suspended or pivoted magnets. Alexander von Humboldt had made such determinations during his expeditions to South America, across Russia and at various places in Europe. Whether the Earth had a magnetic field in the historical or even in the geological past was unknown and also the age of the Earth and of the formations, which documented its geological history, was a matter of speculations. However, it was already known that iron ores, iron meteorites, and rocks visibly containing iron ore minerals such as Magnetite ($Fe_3O_4$) could be strongly magnetic. However, in 1797, von Humboldt discovered a rock unit, which was strongly magnetic without having any visible content of iron ores. The reason for this effect was enigmatic.

During the late 19th and early 20th century it became more and more evident that information about the geomagnetic field in the historical and even geological past could eventually be stored in rocks. Pioneers in this field of research were, for instance, Bernard Brunhes in France, Motonori Matuyama in Japan, and Johann Georg Koenigsberger in Germany. They also found evidence that the geomagnetic field in the past (the so-called paleofield) could have been different from the present field intensity and direction and that the field could eventually also have changed its polarity several times. Based on the work of Koenigsberger, Emile Thellier in Paris developed a reliable method in the 1930s to determine the strength of paleofield at least for historical times using well-dated bricks and pottery from Central Europe and he established curves for the variation of declination, inclination, and field intensity during the last 2000 years. He did not believe that his method could also be applied for rocks. During the same time, Takesi Nagata in Japan made similar experiments and carried out systematic measurements to determine the magnetic properties of minerals, which are possible carriers of information about the paleofield. However, the physics behind this ability of rocks to maintain their magnetism over extremely long times was still cryptic because ordinary handmagnets in use lose their magnetism within relatively short times, and it was not believed by many physicists that rocks would be able to preserve their magnetism over geological times. It was Louis Néel (who later obtained the Nobel Prize for his discoveries) who was able to demonstrate in 1948 that the special property of rocks to store magnetic information over extremely long times is an effect of the small size of its magnetic iron ore particles. His discoveries were also crucial for the development of magnetic tapes in the decades thereafter.

The most frequently occurring magnetic mineral in rocks is Magnetite $Fe_3O_4$, which is called Titanomagnetite when some of the iron is replaced by Titanium. Other important strongly magnetic minerals are the iron oxides Maghemite ($\gamma Fe_2O_3$) and Hematite ($\alpha Fe_2O_3$) and the iron sulfides Pyrrhotite ($Fe_7S_8$) and Greigite ($Fe_3S_4$). Their grain size in rocks varies from $10^{-2}$ to several $10^{+3}$ micrometers. Only grains of magnetite with a size of 0.1 to 1 micrometer are able to carry a remanent magnetization over geological times. Larger or smaller grains are magnetically less stable and can loose their magnetism much faster. Rocks receive their remanent magnetization, which is called their natural remanent magnetization (NRM), during various processes during their geological history. In magmatic rocks such as lavas (e.g., basalt) or intrusions (e.g., granite) the remanence is acquired when cooling in a magnetic field. This remanence is called a thermoremanent magnetization (TRM). In sediments (e.g., limestones and sandstones), the detrital remanent magnetization (DRM) forms during or shortly after the sedimentation when small magnetic particles are oriented in the geomagnetic field and get blocked in the rock matrix. All rock types can also acquire a chemical remanent magnetization (CRM) during processes when the mineralogy of the rock is altered, for instance during a metamorphic overprint. Common to all remanence acquisition processes is that they produce a remanent magnetization parallel to the ambient geomagnetic field and proportional to its intensity. The analyses of the NRM for the determination of the direction of the paleofield require careful laboratory procedures. Even more complicated are the techniques to determine the intensity of the paleofield. It would lead too far and it would be beyond the scope of this book to go into technical details. For this we refer to special textbooks like Dunlop and Özdemir (1997). New insights into the processes of remanence acquisition by sediments can be found in Chap. 3 of this book.

It was not before 1960 or so when geomagnetists had accumulated enough evidence to be convinced that the mean geomagnetic field was that of a geocentric axial dipole and that effects of secular variation are averaged out during several thousands of years. In this case the mean magnetic pole coincides with the geographic north pole (Fig. 1.10). By 1970 it became clear that polarity changes are a common feature in the Earth's history. The polarity changes during the last 150 million years have been documented in detail during the last few decades. They are shown in Fig. 1.11 for the last 5 and in Fig. 1.12 for the last 160 million years. For earlier times polarity changes are also frequently occurring and are a normal behavior of the geomagnetic field; however, their documentation is not yet as perfect as during the last few million years.

We live now in the Brunhes Chron of normal field polarity. The last field reversal was about 780,000 ago. It has been documented in all parts of the world by radiometrically dated lava sequences carrying a TRM and by equally well-dated sedimentary sequences carrying a DRM. These data have been analyzed in Chap. 3 to learn more about the behavior of the geomagnetic field during a reversal and the duration of a reversal process. Since the last reversal about 0.78 million years ago there were several attempts to reverse the field. Such effects are called excursions. The last one

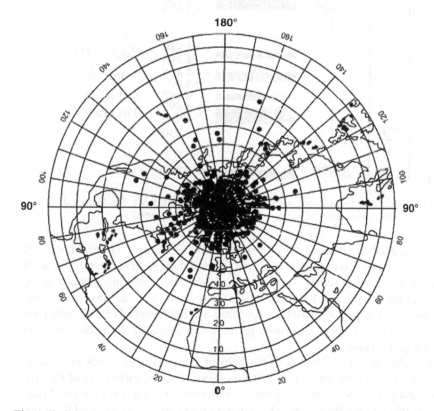

**Fig. 1.10** Geomagnetic poles determined from rocks of the last 20 Ma. They group around the geographic north pole supporting the hypothesis that the time averaged geomagnetic field is that of a geocentric axial pole (after Tarling, 1983)

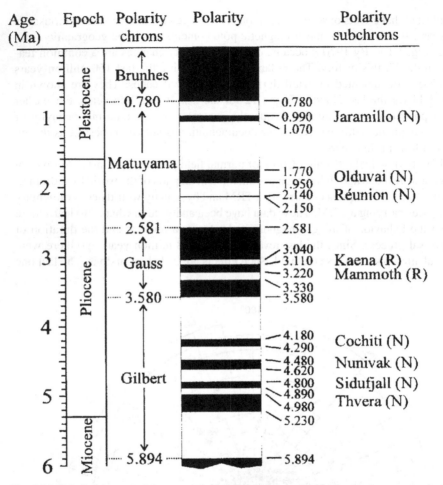

**Fig. 1.11** Polarity changes of the geomagnetic field during the last 5 Ma. *Black*: normal polarity, i.e. todays polarity; *white*: reversed polarity (from Cande and Kent, 1995)

was the so-called Laschamp excursion about 41,000 years ago. The excursions are periods with a duration of several thousand years where the field is relatively weak and seems to reverse in some parts of the globe, whereas in other parts only a strong secular variation is present. Up to now it is still enigmatic by which processes in the Earth core (see Chap. 4) the onset of excursions and field reversals are triggered and why we observe in the Earth's history periods with abundant and periods without any polarity reversals.

It was also discovered that with the help of the paleofield it was possible to prove Alfred Wegeners continental drift hypothesis from the early years of the 20th century and to reconstruct the distribution of the continents in the geological past. However, this will not be discussed here in detail because it is far beyond the scope of this book. There exists abundant literature about this fascinating aspect of geo-magnetism and palaeomagnetism (e.g., Gubbins and Herrero-Bervera, 2007).

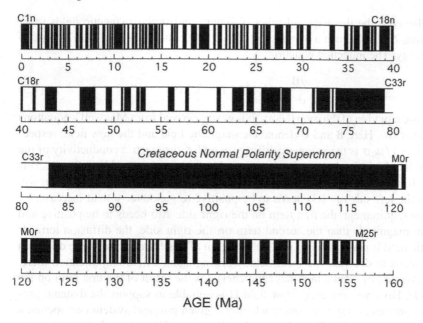

**Fig. 1.12** Polarity changes of the geomagnetic field during the last 160 Ma. While during the Tertiary (the last 65 million years) the polarity changed frequently in a stochastic manner, there was a long period of normal polarity during the Cretaceous (figure courtesy Dennis Kent)

Today there is good evidence that Earth had a magnetic field since at least 3.7 billion years. As rock samples of that age have suffered a series of complex geological, tectonic, and metamorphic processes, the direction of their remanent magnetization is more or less meaningless. Nevertheless, it is a challenge to investigate their magnetic properties and to prove by tests that they still contain some fractions of the original TRM that can be used for paleointensity determinations. Such experiments have recently been made on single crystals of 3.7-billion-years-old rocks, which proved that the Earth had a magnetic field at that time. Measurements of the magnetic field of the Moon and of Mars using satellites showed evidence for the existence of magnetic anomalies, which indicate that these Earth-like bodies also possessed an internal source for a magnetic field at least in their early history. We hope to obtain more information about these fields during the spacecraft missions, which are planned for the coming years. Most recent results are discussed in Chap. 3 of this book.

## 1.4 Basic Dynamo Theory

Today the most widely accepted theory for the explanation of the geomagnetic main field is a dynamo process in the liquid outer core of the Earth. Such a dynamo process has already been proposed by Joseph Larmor in 1919 (Larmor, 1919) and is also believed to be the origin of other planetary and stellar magnetic fields (e.g., Stevenson, 2003). In such a dynamo process, kinetic energy is converted into magnetic energy. The induction equation

$$\frac{\partial \mathbf{B}}{\partial t} = \nabla \times \mathbf{U} \times \mathbf{B} + \lambda \triangle \mathbf{B}, \tag{1.3}$$

is the basic equation of dynamo theory. It is easily derived from Maxwell's equations and Ohm's law. Here $\mathbf{B}$ and $\mathbf{U}$ denote the magnetic field and the flow field, respectively; $\lambda = 1/\mu_0\sigma$ is the magnetic diffusivity and $\sigma$ the electric conductivity of the medium in which the dynamo is operating. The interpretation of Equation (1.3) is straight forward: $\mathbf{B}$ is the magnetic field to be generated and amplified. The growth term on the left side of (1.3) needs to be positive to guarantee a field increase. To meet this requirement, the first term on the right side also needs to be positive and larger in magnitude than the second term on the right side, the diffusion term. If magnetic field is generated in a particular region of the dynamo, magnetic diffusion always tends to decrease the local field, working against field amplification there. The conversion of kinetic into magnetic energy is the result of the first term on the right side. However, not every flow field $\mathbf{U}$ is suitable to support the dynamo process. A convenient way to estimate whether a given physical system can operate a dynamo is the magnetic Reynolds number $R_m = \mu_0\sigma UL$, where $L$ denotes a typical spatial scale of the system. For a dynamo to work, one requires $R_m >> 1$. If this condition is not fulfilled, magnetic diffusion very effectively acts against field amplification.

This short discussion of the induction equation (1.3) indicates that three conditions need to be fulfilled to operate a dynamo: one needs a systems with large electric conductivity, rapid fluid motion, and a sufficiently large spatial scale. The Earth's outer fluid core is such a regime and thus thought to be the region where the geomagnetic field is generated. From geomagnetic induction studies the conductivity is estimated at $\sigma \approx 5 \times 10^5$ S/m (Parkinson and Hutton, 1989), while the flow velocity can be estimated from the westward drift of the geomagnetic field at $U \approx 1$ m/s. With $L = 2,000$ km, the radial extent of the outer core, the terrestrial Reynolds number is of the order of 2,000, large enough to argue in favor of a dynamo process operating in the Earth's deep interior.

A simple realization of a dynamo is the $\alpha - \Omega$ dynamo (Parker, 1970). This kinematic dynamo assumes a differentially rotating dynamo regime in which a poloidal seed field $\mathbf{B_P}$ (see Fig. 1.13) is partially converted into a toroidal field $\mathbf{B_T}$ because of differential rotation of the core. This conversion is called $\Omega$-effect. Although the generation of toroidal magnetic fields represents a magnetic field generation process, differential rotation alone does not represent a dynamo process, as the poloidal seed field needs to be amplified. Thus, a process is required to convert a toroidal magnetic field into a poloidal field. Only if the complete cycle $\mathbf{B_P} \Rightarrow \mathbf{B_T} \Rightarrow \mathbf{B_P}$ can be realized a dynamo process is realized. In a rotating system Coriolis forces $\mathbf{F_{cor}} = 2\mathbf{\Omega} \times \mathbf{U}$, where $\mathbf{\Omega}$ denotes the rotation vector, need to be taken into account. In its simplest form the $\alpha$-effect describes the twist of a toroidal field into a poloidal one due to the Coriolis forces (see Fig. 1.13). The $\alpha$-effect therefore completes the loop initiated by the $\Omega$-effect.

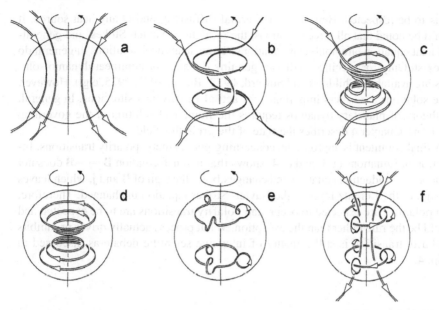

**Fig. 1.13** Magnetic field amplification in the $\alpha - \Omega$ dynamo. The dynamo cycle starts with poloidal field lines (**a**) whirled up by differential rotation into toroidal field lines (**c** and **d**). Toroidal field lines (**d**) are distorted by cyclonic motions, producing poloidal field lines (**e** and **f**) amplifying the primary poloidal field in a dynamo (courtesy by J.J. Love; see also Love, 1999)

The $\alpha - \Omega$ dynamo is a kinematic dynamo model, in that only the action of the fluid flow on the magnetic field is taken into account. Any modification of the flow system due to the generated magnetic field is neglected. To overcome this simplification, the dynamic coupling between the magnetic field and the flow field needs to be studied. The Navier–Stokes equation

$$\rho \left( \frac{\partial \mathbf{U}}{\partial t} + (\mathbf{U} \cdot \nabla)\mathbf{U} \right) = -\nabla p - 2\rho \mathbf{\Omega} \times \mathbf{U} - \rho \nu \nabla^2 \mathbf{U} + \rho \alpha T \mathbf{g} + \mathbf{j} \times \mathbf{B} \qquad (1.4)$$

governs the dynamic behavior of the fluid. The terms on the right-hand side describe the influence of pressure gradients, the Coriolis forces, viscosity, buoyancy, and Lorentz forces, respectively, on the flow. Here $\rho$ is the mass density (assumed constant), $p$ the pressure, $\mathbf{j}$ the electric current density, $\nu$ the kinematic viscosity, $\mathbf{g}$ gravity, $\alpha$ the coefficient of thermal expansion, and $T$ the temperature. It is in particular the Lorentz force term that couples the electromagnetic field with the flow field.

As also the temperature $T$ is involved, a further generalized heat equation

$$\frac{\partial T}{\partial t} = \kappa \nabla^2 T - \mathbf{U} \cdot \nabla T + Q \qquad (1.5)$$

needs to be respected. Here $\kappa$ is the thermal diffusivity and $Q$ any heat source. It should be noted that all three equations, the induction, Navier–Stokes, and generalized heat equation, are nonlinear equations. Thus, analytical solutions in general do not exist. Only for specific conditions genuine dynamic or nonlinear dynamo solutions are available (Childress and Soward, 1972; Busse, 1973, 1975, e.g.). However, these solutions are of less importance for actual geophysical situations. In general, the theory of planetary dynamos requires the numerical solution of the governing equations. Chapter 4 describes the state of the art in this field.

A final comment is appropriate, concerning geomagnetic polarity transitions. Inspection of Equations (1.3) and (1.4) shows that a transformation $\mathbf{B} \rightarrow -\mathbf{B}$ does not influence the induction equation and changes both, the sign of $\mathbf{B}$ and $\mathbf{j}$, which leaves the sign of the Lorentz force in the Navier–Stokes equation unchanged. Therefore, both polarities are expected to occur, and polarity transitions are to be expected and should be the rule rather than the exception. What process actually drives or inhibits a polarity transition is still a matter of intensive scientific debate as discussed in Chap. 4.

## 1.5 Earth and Space

Space between the planets is not empty, but filled with a fast streaming, magnetized plasma of solar origin. This solar wind is a supermagnetosonic flow of mainly electrons and protons in which the Earth and other bodies of our solar system are embedded. Its mean particle density and flow velocity in the ecliptic plane is about $8 \ \mathrm{cm}^{-3}$ and 470 km/s, respectively. Next to protons, Helium ions are the dominant particles with an abundance of a few percent. Flow velocities of up to 2,000 km/s have been observed during so-called coronal mass ejections (CME), explosions in the corona of the Sun that spew out solar particles into the preexisting solar wind. The normal solar wind causes a solar mass loss of about 1,000 kg/s, while during a CME, a total of $10^{13}$ kg are ejected. The solar wind electrons and protons exhibit kinetic temperatures of the order of $10^5 K$ and more, which corresponds to mean particle energies of 10–100 eV. Particles with much higher energies, up to the GeV level, are also present. They also originate in the Sun or are cosmic particles accelerated elsewhere.

With the solar wind the solar magnetic field is carried out into space. Due to its extremely large electrical conductivity, solar wind plasma and solar magnetic field are tightly coupled, which means that the magnetic field is frozen into the plasma flow. At Earth orbit the mean magnetic field strength is of the order of 7 nT.

Every planetary body is embedded in the solar wind, filling the interplanetary space. Around every such body an interaction region develops in order to accomplish the solar wind flow around this body. The actual interaction region depends very much on the nature of the planetary body. The moon of the Earth, without an atmosphere and a strong global magnetic field, is an example where the solar wind impinges unimpeded on the surface. Active comets with their strong outgassing surfaces manage to prevent the solar wind reaching their surface. A magnetic cavity forms, acting as a shield against the energetic solar wind particles and filled only with a gas of cometary origin.

Earth with its strong dynamo generated magnetic field develops a special interaction region, the magnetosphere. The solar wind speed is usually larger than the characteristic velocity of magnetosonic waves in the solar wind plasma, that is the magnetosonic Mach number is larger than 1. To accomplish the flow around Earth, a standing bow shock wave develops upstream of the Earth. In this bow shock with a typical stand-off distance of $R_{BS} = 13\ R_E$, where $R_E = 6,371$ km is the Earth radius, kinetic flow energy of the solar wind particles is converted into thermal energy up to an extent that a sub-magnetosonic situation emerges. Downstream of this bow shock the actual flow around the Earth is accomplished in the so-called magnetosheath region. The plasma of the magnetosheath, however, cannot penetrate the magnetosphere proper region, but is separated from it by the magnetopause. Due to the action of Lorentz forces the solar wind electrons and protons cannot penetrate close to the Earth, but a plasma boundary is formed as the boundary between the solar wind plasma and the magnetospheric plasma; this boundary is called magnetopause (Fig. 1.14).

The magnetopause is a boundary where the solar wind dynamic pressure, $p_{SW} = 2\rho v_{SW}^2$, and the magnetic pressure of Earth's magnetic field, $p_{mag} = B^2/2\mu_0$, are in equilibrium. Here, $\rho$ denotes mass density, and $v_{SW}$ the solar wind bulk velocity.

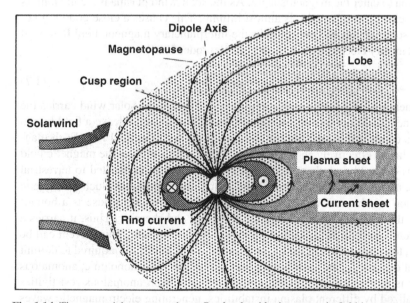

**Fig. 1.14** The terrestrial magnetosphere. On the dayside the magnetic field is compressed its interaction with the solar wind; on the nightside it is stretched out to form the magnetotail with its lobe and plasma sheet regions. The stretched configuration is supported by the tail current. In the magnetosphere proper trapped particles constitute the ring current (modified after Baumjohann, 1991)

On the stagnation flow line the magnetopause distance, $R_{MP}$, is given as (e.g. Kallenrode, 2004):

$$R_{MP} = \sqrt[6]{\frac{2B_E^2}{\mu_0 \cdot p_{SW}}} \cdot R_E, \tag{1.6}$$

where $B_E$ denotes the terrestrial magnetic field strength at the equatorial surface. For nominal solar wind conditions the magnetopause distance is about 10 $R_E$. The region downstream of this magnetopause is called the magnetosphere. It is filled with plasma of ionospheric and solar wind origin. The dynamical behavior of the plasma protons and electrons is governed by the terrestrial magnetic field, thus the name magnetosphere.

In extreme cases, when the solar wind density and speed are very large, the magnetopause can move in to about 5 $R_E$. For current magnetic field conditions it never reaches the Earth's surface. Therefore, the magnetosphere may be viewed at as a natural shield against solar wind particles.

However, there are three possibilities for solar wind plasma and particles to enter the magnetosphere proper and interact with the atmosphere and even surface. First, plasma entry is possible in the cusp regions of the magnetosphere where the Earth magnetic field is about normal to the magnetopause. Here, particles can enter along magnetic field lines with the Lorentz force being negligible (Fig. 1.14). Second, a process called magnetic reconnection provides another possibility for solar wind plasma to enter the magnetosphere. As the solar wind plasma is a collisionless plasma its electric conductivity is almost infinite. This causes a close coupling between the solar wind bulk flow velocity, the interplanetary magnetic field $\mathbf{B}_{SW}$, and the interplanetary electric field $\mathbf{E}_{SW}$ (e.g. Kallenrode, 2004):

$$\mathbf{E}_{SW} = -\mathbf{v}_{SW} \times \mathbf{B}_{SW}. \tag{1.7}$$

A consequence of this frozen-in theorem is not only that the solar wind carries the solar magnetic field out into space, but also that plasma cannot leave a field line to which it is connected, as the plasma always moves with the field line. Each magnetic field line is effectively an electric equipotential line. Only if the magnetic field topology is changed such that interplanetary field lines are connected to terrestrial field lines, plasma transport across the magnetopause is possible. Such a magnetic reconnection is schematically shown in Fig. 1.15. The magnetopause is a boundary where the magnetic field is suddenly changing and shearing. Thus, it carries a strong electric current, the so-called Chapman-Ferraro current. This current can be very large, too large to be carried by the plasma. A mechanism is required to delimit the current density. As the solar wind plasma is collisionless nonohmic, anomalous resistivity is required to control the current density. Such anomalous resistivities can be realized by different plasma instabilities, generating electromagnetic waves acting as scatterers for the plasma particles. If the Chapman-Ferraro current is limited by such processes it can be viewed as switching on a counter-current in the magnetopause. This counter-current generates its own magnetic field, which leads to a change in the overall–magnetic field topology. A common way to visualize this

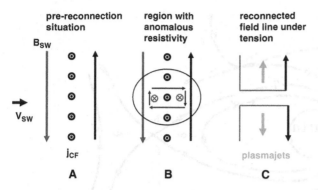

**Fig. 1.15** Magnetic reconnection: (**A**) due to antiparallel magnetic fields a sheet current exists; (**B**) increasing this sheet current leads to anomalous resistivity and the switch-on of a counter current; (**C**) due to the magnetic field change caused by the electric current modification, the field lines "reconnect" and plasmajets are generated

topology change is to break solar wind field and terrestrial field lines and reconnect them in the way depicted in Fig. 1.15. The newly reconnected field lines are strongly bent and the associated magnetic tension leads to rapid plasma transport out of the magnetic reconnection region. Magnetic reconnection can also be viewed as an antidynamo process, as magnetic energy of the pre-reconnection situation is converted into kinetic energy of the plasma jets.

The global effect of this reconnection process is plasma and magnetic flux transport from the dayside of the magnetosphere down to the nightside; the magnetotail forms in this way. Continuous erosion of magnetic flux at the dayside is not possible. A process is needed to balance magnetic flux transport. Again, magnetic reconnection is the physical process, which guarantees such a balance via reconnection of field lines in the tail region of the magnetosphere (see Fig. 1.16). Both, reconnection on the dayside and the nightside, cause a very dynamic magnetosphere with strong plasma transport and convection, polar lights, and rapid geomagnetic field variations.

The third process by which plasma particles can enter the magnetosphere is direct entry of energetic particles. This process is important if the gyroradius of the particle is comparable to the scale of the magnetosphere. Figure 1.17 illustrates possible particle behavior. Low energy particles entering at high geomagnetic latitudes are reflected at their mirror points, while the more energetic ones can penetrate deep into the atmosphere. This is also true for very energetic particles entering closer to the magnetic equator. Other particles are trapped in the geomagnetic field and constitute the so-called radiation belts or van Allen Belts. These trapped particles also carry the ring current depicted in Fig. 1.14. Only particles with extremely high energies, TeVs, can penetrate through the atmosphere and reach the ground.

The magnetosphere and the geomagnetic field are often regarded as a shield against energetic particles. However, it is actually the atmosphere that inhibits energetic particles to reach the Earth's surface and damage the biosphere. The magnetosphere and the geomagnetic field play more the role of moderators

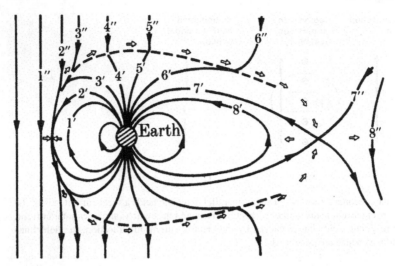

**Fig. 1.16** Recycling of magnetic flux tubes in the open magnetosphere. The field line numbered 1 is an unperturbed solar wind magnetic field line, which opens via magnetic reconnection (2) and is transported toward the magnetotail (3 to 6), where reconnection occurs again transporting field lines (7 and 8) back to the dayside and out down the magnetotail. A prime denotes the former terrestrial parts of a field line, a double prime former solar wind parts (after Levy et al., 1964)

controlling where particles can penetrate deeper into the atmosphere. For current geomagnetic conditions the regions around the magnetic poles in the northern and southern hemisphere are the most vulnerable places. There, not only the majestic polar lights are generated (e.g. Eather, 1980), but enhanced radiation can even be measured in airplanes flying the polar routes (e.g. Wissmann, 2006). Also, the region of the South Atlantic Anomaly is more vulnerable to energetic particles than anywhere else as particle trapping in the geomagnetic field is less effective there. Especially, spacecrafts are endangered in this regime and effected by severely increased numbers of so-called single event upsets, perturbations of their electronics (e.g. Ziegler and Lanford, 1979; Harboe-Sorensen et al., 1990).

With man populating space, using artificial satellites and technologies with ever increasing complexity, geomagnetic field variations caused by the dynamical processes in the Earth magnetosphere also gained economic interest. For example, during geomagnetic storms, geomagnetic variations with changes of the field strength of up to 2,000 nT, electric power supply can be disrupted as happened in Quebec in 1989 (e.g., Kappenman, 1999). A whole new discipline, space weather research, has developed in trying to predict such geomagnetic storms (e.g., Bothmer and Daglis, 2006).

The most drastic geomagnetic variation is the polarity transition. The final Chap. 5 is devoted to the situation during a field reversal when the geomagnetic field is weak and of complex topology. Computer simulations show that in these situations considerable effects on the magnetosphere and its property as a shield for the Earth against radiation need to be considered. It could be shown that also during the

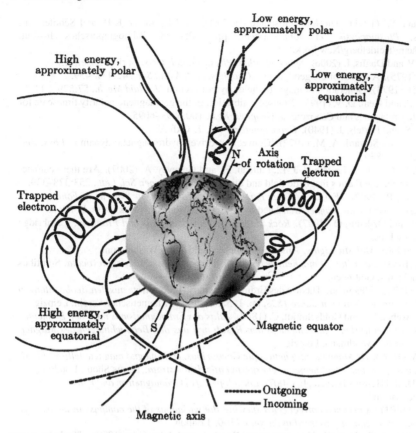

**Fig. 1.17** Energetic particles and the geomagnetic field. *Red* areas denote the geomagnetic pole regions with their large field strengths; the *blue* regime is the south Atlantic anomaly (modified after Alonso and Finn, 1972).

critical phases of low geomagnetic field intensity the Earth does not loose its magnetosphere. However, the magnetospheric structure becomes much more complex. Also, the magnetopause distance from the surface of the Earth is reduced and so is its property as a shield for system Earth. Radiation from within and from outside of our solar system can penetrate more easily into lower layers of the atmosphere with various effects, for instance, on the ozone layer and eventually also on life. Chapter 5 is discussing these effects in detail.

# References

Alonso, M. and Finn, E. (1972). *Fundamentals of University Physics, Vol. 2.* Addison Wesley, Reading.
Balmer, H. (1956). *Beiträge zur Geschichte der Erkenntnis des Erdmagnetismus.* Aarau.

Baumjohann, W. (1991). Die Magnetosphäre der Erde. In Glassmeier, K.H. and Scholer, M., editors, *Plasmaphysik im Sonnensystem*, pp. 105–118. Bibliographisches Institut, Mannheim/Heidelberg/New York.

Bothmer, V. and Daglis, I. (2006). *Space Weather. Physics and Effects*. Springer, Berlin.

Busse, F. (1975). A model of the geodynamo. *Geophys. J. R. Astro. Soc.*, 42:437–459.

Busse, F. H. (1973). Generation of magnetic fields by convection. *J. Fluid Mech.*, 57:529–544.

Cande, S. C. and Kent, D. V. (1995). Revised calibration of the geomagnetic polarity timescale for the Late Cretaceous and Cenozoic. *J. Geophys. Res.*, 100:6093–6095.

Chapman, S. and Bartels, J. (1940). *Geomagnetism, Vol. 1*. Oxford.

Childress, S. and Soward, A. M. (1972). Convection-driven hydromagnetic dynamo. *Phys. Rev. Lett.*, 29:837–839.

Courtillot, V., Gallet, Y., Le Mouël, J.-L., Fluteau, F., and Genevey, A. (2007). Are there connections between the Earth's magnetic field and climate? *Earth Planet. Sci. Lett.*, 253:328–339.

de Maricourt, P. (1902). *Epistle of Peter Peregrinus of Maricourt, to Sygerus of Foncaucourt, soldier, concerning the magnet,* translated by S. P. Thompson. Chiswick Press, London.

Dunlop, D. and Özdemir, O. (1997). *Rock Magnetism: Fundamentals and Frontiers*. Cambridge University Press.

Eather, R. (1980). *Majestic Lights*. American Geophysical Union, Washington.

Gauss, C. (1833). *Intensitas vis magneticae terrestris ad mensuram absolutam revocata*. Sumtibus Dieterichianis, Göttingen.

Gauss, C. (1839). Allgemeine Theorie des Erdmagnetismus. In *Resultate aus den Beobachtungen des Magnetischen Verein im Jahre 1838*, pp. 1–52. Göttinger Magnetischer Verein, Leipzig.

Gauss, C., Weber, W., and Goldschmidt, C. (1840). *Atlas des Erdmagnetismus nach den Elementen der Theorie entworfen: Supplement zu den Resultaten aus den Beobachtungen des Magnetischen Vereins*. Weidmann, Leipzig.

Gilbert, W. (1600). *De Magnete, Magneticisque Corporibus, et de magno magnete tellure; Physiologia nova, plurimis et arguementis et experimentis demonstrata*. Petrus Shout, London .

Gubbins, D. and Herrero-Bervera, E. (2007). *Encyclopedia of Geomagnetism and Paleomagnetism*. Springer, Berlin.

Halley, E. (1701). *A new and correct chart shewing the variations of the compass in the western and southern oceans as observed in the year 1700*. London.

Harboe-Sorensen, R., Daly, E., Underwood, C., Ward, J., and Adams, L. (1990). The behaviour of measured SEU at low altitude during periods of high solar activity. *IEEE Trans. Nucl. Sci.*, 37:1076–1083.

Haug, G., Günther, D., Peterson, L., Sigman, D., Hughen, K., and Aeschlimann, B. (2003). Climate and the collapse of Maya civilzation. *Science*, 299:1731–1735.

IAGA (2005). The 10th generation international geomagnetic reference field. *Phys. Earth Planet. Int.*, 151:320–322.

Kallenrode, M.-B. (2004). *Space Physics: An Introduction to Plasmas and Particles in the Heliosphere and Magnetospheres*. Springer, Berlin/Heidelberg/New York.

Kappenman, J. (1999). Geomagnetic storm and power system impacts: advanced stormforecasting for transmission system operations. *IEEE*, 2:1187 – 1191.

Kertz, W. (1999). *Geschichte der Geophysik,* edited by K.H. Glassmeier and R. Kertz. G. Olms Verlag, Hildesheim.

Larmor, J. (1919). How could a rotating body such as the sun become a magnet? *Rep. Brit. Assoc.*, 87:159–160.

Levy, R., Petschek, H., and Siscoe, G. (1964). Aerodynamic aspects of the magnetospheric flow. *A.I.A.A. Jl.*, 2:2065–2076.

Love, J. (1999). Reversals and excursions of the geodynamo. *Astron. Geophys.*, 40:14–19.

McDonald, K. L. and Gunst, R. H. (1968). Recent trends in the Earth's magnetic field. *J. Geophys. Res.*, 73:2057–2067.

Palmer, S. J., Rycroft, M. J., and Cermack, M. (2006). Solar and geomagnetic activity, extremely low frequency magnetic and electric fields and human health at the Earth's surface. *Surv. Geophys.*, 27:557–595.

Parker, E. (1970). The generation of magnetic fields in astrophysical bodies. 1. the dynamo equations. *Astrophys. J.*, 162:665–673.

Parkinson, W. D. and Hutton, V. R. S. (1989). The electrical conductivity of the earth. In Jacobs, J., editor, *Geomagnetism, Vol. 3*, pp. 261–321. Academic Press, London.

Stevenson, D. (2003). Planetary magnetic fields. *Earth Planet. Sci. Lett.*, 208:1–11.

Tarling, D. (1983). *Palaeomagnetism: Principles and applications in geology, geophysics and archaeology*. Chapman and Hall, London.

Wang, Y., Y. Pan, S. Parsons, M. Walker, S. Zhang (2007). Bats respond to polarity of a magnetic field. *Proc. R. Soc. B*, 274:2901–2905.

Winklhofer, M. (2004). Vom magnetischen Bakterium zur Brieftaube. *Phys. Unserer Zeit*, 35:120–127.

Wissmann, F. (2006). Long-term measurements of h*(10) at aviation altitudes in the northern hemisphere. *Radiat. Prot. Dosim.*, 121:347–357.

Wiltschko, W. and R. Wiltschko (2005). Magnetic orientation and magnetoreception in birds and other animals. *J. Comp. Physiol.*, 191:675–693.

Ziegler, J. and Lanford, W. (1979). Effect of cosmic rays on computer memories. *Science*, 206:776–788.

# Chapter 2
# The Recent Geomagnetic Field and its Variations

Hermann Lühr, Monika Korte and Mioara Mandea

## 2.1 Introduction

The Earth's magnetic field is the sum of several contributions. These are internal ones, i.e. the core field, also known as the main field, and the crustal or lithospheric field, and external ones originating from ionospheric and magnetospheric currents. The core field is by far the most prominent part. The lithospheric field has its origin in remanent and induced magnetization of the crust and upper mantle. Due to its local and regional variability, the individual features are referred to as magnetic anomalies. Strongly magnetized material or large geological bodies create some outstanding anomalies. A well-known example is the Kursk anomaly, in Ukraine, with peak amplitudes of some 3,000 nT in aeromagnetic data at 500 m altitude, and extending over several 100 km. Weaker anomalies of a few tens to hundreds of nT are present almost everywhere on Earth.

These internal parts of the geomagnetic field can well be described mathematically by a spherical harmonic expansion . This is the representation of the magnetic field potential as a series of multipoles: $n = 1$ represents the dipole contribution, $n = 2$ that from a quadrupole, $n = 3$ from an octupole and so on (see Sect. 2.3.1 for more details). The degrees thus are a measure of spatial wavelength and the field can be analyzed by plotting the power, the squared magnetic field strength, $B^2$, averaged over the globe as a function of spherical harmonic degree. Figure 2.1 shows such a spatial spectrum of the recent magnetic field power at the Earth's surface. The approximate corresponding wavelength can be calculated by dividing the circumference of the Earth (ca. 40,000 km) by the given spherical harmonic degree.

Such a spectrum shows two branches with very different slopes. On a logarithmic scale the two parts can reasonably well be approximated by linear fits if the dipole is excluded. The transition from one slope to the other occurs between degrees 13 and 15.

Hermann Lühr
GeoForschungsZentrum Potsdam Telegrafenberg 14473 Potsdam Germany

Monika Korte
GeoForschungsZentrum Potsdam Telegrafenberg 14473 Potsdam Germany

Mioara Mandea
GeoForschungsZentrum Potsdam Telegrafenberg 14473 Potsdam Germany

K.-H. Glaßmeier et al. (eds.), *Geomagnetic Field Variations*, Advances in Geophysical and Environmental Mechanics and Mathematics,
© Springer-Verlag Berlin Heidelberg 2009

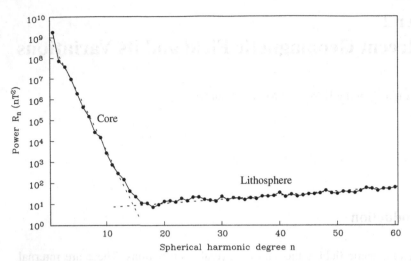

**Fig. 2.1** Main and lithospheric field spatial spectrum, $R_n = (n+1)\sum_{m=0}^{n}((g_n^m)^2 + (h_n^m)^2)$ with the notation of Equation 2.4, at the Earth's surface. Spatial wavelength in kilometre is obtained by dividing the circumference of the Earth by spherical harmonic degree $n$

At smaller degrees the spectrum is dominated by the field generated in the core. The dipole (degree 1) is quite outstanding. The power of the degrees 2 to about 13 falls off with a slope that is consistent with a source at about 3,000 km below the surface. This distance coincides reasonably well with the depth where the top of the core is. The nearly flat spectrum at degrees beyond 15 is an indication for the short distance to the field sources and is dominated by the lithospheric field. Approximately between degrees 13 and 15, or 3,000 to 2,600 km wavelength, the observed field is strongly influenced by both core and crustal field contributions. The short wavelengths of the core field are masked by the crustal signal, and the long wavelength features of the crustal field are buried under the large core field amplitudes.

The space around the Earth, into which the geomagnetic field extends, is called the magnetosphere . It extends to about 10 Earth radii ($R_E$) at the sub-solar point and is pulled out to several 100 $R_E$ on the night side by the solar wind. This comet tail-like structure has a radius of about 20 $R_E$. The shape of the magnetosphere is determined by the pressure balance between the solar wind and the geomagnetic field. Generally, impinging solar wind particles cannot penetrate the magnetosphere. However, at certain interplanetary magnetic field (IMF) orientations they can enter and drive magnetospheric currents. A variety of different current systems exists. Some of them extend down to the polar ionosphere, others form a ring around the Earth in the equatorial plane. All of these magnetospheric currents can produce highly variable magnetic fields.

The altitude range of the ionosphere is typically allocated between 90 and 1 000 km above the Earth. It consists of a mixture of charged and neutral particles. The dynamics of the charged particles is strongly dependent on collisions with the neutral ones. Best conditions for horizontal electric currents are found around 110 km altitude in the sunlit part of the atmosphere. Large scale ionospheric currents at middle latitudes generate diurnally varying magnetic fields with amplitudes up to 80 nT at the Earth's surface.

It is well know that the Earth's magnetic field at all times is subject to temporal variations. They cover a huge range of timescales, from fractions of seconds to geological time spans. The short period variations, from seconds to days, are primarily of external origin. They are mostly caused by influences from solar activity and solar variability on magnetosphere and ionosphere, e.g. by coronal mass ejections, coronal hole streams, solar wind sector boundaries, and IMF orientation changes, or by rather regular variations in atmospheric heating and ionospheric conductivity with solar irradiation. All these varying external fields additionally induce electric currents within the Earth, which in turn produce varying induced magnetic fields depending on the electrical conductivity of the subsurface layers. Other kinds of secondary, induced fields varying at short times are produced in the oceans. The movement of electrically conducting sea water through the geomagnetic field generates secondary magnetic fields through the magneto-hydrodynamic process of motional induction.

The very long period variations, from decadal up to polarity reversals that occur on million year time scales, are caused by changes in fluid flow as part of the geodynamo process in the Earth's outer core. These core field changes are named secular variation. Most difficult to study are the intermediate variations, from years to decades, where influences from external sources and secular variation overlap. However, a particularly interesting feature of secular variation occurs just on the shortest core field variation time-scales that are observable after propagation through the weakly conducting mantle: the so-called geomagnetic impulses or jerks. They are observed as sharp changes of trend in secular variation within one or two years. Due to the rapidity of the changes they are useful for studying core processes and parameters by providing some specific constraints.

In this chapter we report on the progress made over the last few years towards a better understanding of the recent field and secular variation, including geomagnetic jerks. Obtaining good data from observations is a prerequisite, and in Sect. 2.2, modern observatory and satellite measurements are described. Global geomagnetic field models based on spherical harmonic analyses are powerful tools to study characteristics of the field and its variations. A variety of recent models is presented in Sect. 2.3 together with the results and conclusions obtained from them. This includes details of the present field configuration, secular variation from geomagnetic jerks to the dipole behavior over a few thousand years, and global descriptions of the lithospheric field. Separating the different field contributions is a persisting problem in any recent field studies, including global modelling approaches. Only an improved understanding of the different external field sources can lead to improved core field models. This is the topic of Sect. 2.4. An outlook to expected future results and unresolved questions is given in the final section.

## 2.2 Recording the Magnetic Field

### 2.2.1 Geomagnetic Observatories

The geomagnetic observatories can be regarded as the backbone for recording the geomagnetic field. They have a long-standing tradition in accurate measurements of three field components, although prior to 1833 measurements were of direction (declination and inclination) only. Gauss did not only introduce the representation of the magnetic field by spherical harmonic functions, but he also developed a method to determine the field intensity absolutely (Gauss, 1833, 1839). His approach, which was accepted worldwide, helped to obtain high quality magnetic field readings, independent of instrument type, temperature, and time, from then on. Today, magnetic field variations are measured by electronic transducers rather than mechanically suspended magnets, and the definition of the magnetic field strength is based on the gyro-magnetic ratio of the proton Larmor frequency rather than on the force a magnet exerts (Jankowski and Sucksdorff, 1996). The idea of a self-consistent check of all three field components, however, is still the same. For the continuous recording of the field, vector fluxgate magnetometers are commonly used. However, these instruments may suffer from drifts and show dependencies on environmental influences. Proton precession magnetometers or optically pumped instruments can provide readings of the magnetic field magnitude with absolute accuracy. Measurements of these two instruments have to be combined in a suitable way for deriving calibrated vector data. Furthermore, vector data have to be described in a well-defined coordinate system. Finding the direction of the geomagnetic field in a geographically oriented frame is a demanding task. An error of 5 arc seconds in inclination is, for example, equivalent to a 1 nT deviation. The accuracy of orientation measurement is thus most critical. At geomagnetic observatories the orientation today is generally determined by nonmagnetic theodolites equipped with a fluxgate sensor to determine declination and inclination absolutely.

The number of magnetic observatories reporting high-quality data has been increasing in general in recent years (see Fig. 2.2b). This is largely due to the initiative of the international INTERMAGNET programme. Only those observatories, which maintain a certain level of data quality and can provide near-real time (preliminary) data, are accepted for programme membership. There is a promising development that more and more observatories try to meet these standards. Presently the number of INTERMAGNET observatories amounts to 106, and their distribution is shown in Fig. 2.2. In spite of the growing number of observatories, vast regions, particularly in the southern hemisphere, are not covered. This imposes severe limitations for the global modeling of the geomagnetic field even though the quality of measurements has improved impressively.

(a)

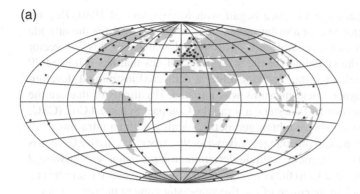

(b)

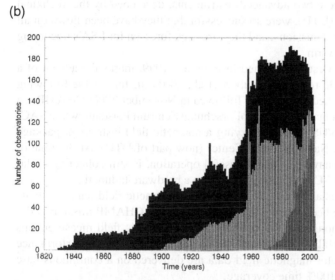

**Fig. 2.2** (a) Map of INTERMAGNET observatories in 2007. (b) Number of observatory data by time as held at the World Data Centers for Geomagnetism. *Black* are annual mean values, *light and dark gray* give the number of observatories for which hourly means and minute values, respectively, are digitally available for a year

## 2.2.2 Satellite Missions

A good spatial coverage of magnetic field measurements on global scale can be achieved by low flying satellites in near polar orbits. The series of magnetic field survey satellites started with the Cosmos 49 satellite in 1964. This short one-month mission was followed by the POGO-2,4,6 series lasting from 1965 to 1971. All these satellites measured only the total field strength. It soon became clear that good global magnetic field models cannot be derived from scalar data alone. The era of vector field measurements from space began with Magsat around 1980. Key instruments were a vector and an absolute scalar magnetometer, and for the attitude determination a star sensor was employed. The approach realized by this first vector magnetic mission is the guideline for the series of satellites that has followed 20 years later. In preparation for the Ørsted satellite a dedicated instrument development program for magnetic field missions was initiated. Highlights of that are the new fluxgate magnetometer and the star camera. The Compact Spherical Coil (CSC) sensor for the fluxgate combines a high geometric stability with an ultra-linear response to ambient magnetic field changes. Likewise, the Advanced Stellar Compass (ASC) combines high-performance attitude determination precision with advanced processing techniques. Thanks to the nonmagnetic design of the new camera head a closely spaced, optimal arrangement of the fluxgate—star camera instrument package can be realized. These two advanced instruments, developed by the Technical University of Denmark (DTU), were so successful that they have been flown on all subsequent magnetic field missions, and they are even foreseen for ESA's upcoming constellation mission Swarm.

The Danish satellite Ørsted, launched in February 1999, marks the advent of a new era in magnetic field research (Neubert et al., 2001). In July of the following year CHAMP was launched, and SAC-C followed in November 2000. CHAMP is a national German project managed by GeoForschungsZentrum Potsdam, while SAC-C is a joint Argentinian/US mission carrying a magnetic field instrument package provided by the Danish National Space Center (now part of DTU). At the time of writing, Ørsted, after many years of successful operation, is still collecting scalar magnetic field data. SAC-C suffered early on from a hardware failure that prevented the retrieval of vector data. Nevertheless, the scalar magnetic field data, available until end of 2004, have been used in various studies. The CHAMP mission is described here in some more detail as an example for magnetic field measurements from space. Up to now, all its essential systems are operating to full performance and both scalar and vector magnetic field data are delivered on a continuous base providing practically a 100% time coverage.

The CHAMP satellite is carrying a variety of different instruments, which serve the purposes of the three mission objectives: study of the Earth's gravity and magnetic fields and sounding of the atmosphere (Reigber et al., 2002). Figure 2.3 shows the arrangement of instruments on the spacecraft. The Overhauser scalar and three-component fluxgate magnetic field sensors are placed on a 4-m long boom. This helps to keep the magnetic disturbances from the spacecraft on a low level at the magnetometers positions. At the center of gravity the triaxial accelerometer is accommodated. The prime purpose of this instrument is to serve the gravity studies. Slightly below the satellite body the Digital Ion Drift-Meter (DIDM) is mounted. This instrument is measuring the dynamics of the charged particles in the ionosphere. Last but not least, there is the GPS receiver serving several purposes. First, it gives the position of the spacecraft; second, it provides the time with a microsecond precision and offers a reference frequency; third, it can be used to perform radio occultation for atmospheric sounding.

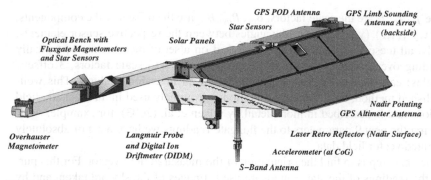

Fig. 2.3 CHAMP Instrumentation

## 2.2.3 Satellite Data Processing

Here we briefly outline which instruments on CHAMP contribute to the retrieval of precise magnetic field data and how the readings are preprocessed. It has to be noted that the approach described is nowadays a kind of standard applied in all magnetic field mapping missions.

The CHAMP Overhauser magnetometer samples the field strength at a rate of 1 Hz with a resolution of 10 pT. Vector field measurements are performed at a rate of 50 Hz with a resolution of 0.1 nT. Since we want to track the variations of the geomagnetic field in all details, it is important that instrumental drifts do not mimic secular variation. For that reason a dedicated calibration approach, applicable in orbit, was implemented. The underlying idea is to trace back all measured quantities to well-documented standards.

Starting with the Overhauser readings, the Larmor frequency at the output of this instrument is directly proportional to the ambient field strength. The proportionality factor, the gyro-magnetic ratio, is a well characterized number maintained as one of the prime physical standards. The GPS receiver provides the precise time gate for counting the oscillations in the Overhauser instrument. GPS clocks are known to be well maintained to nanosecond precision. Under these conditions it is valid to use the term "absolute scalar field data" for the magnitude measurements.

The readings of a fluxgate magnetometer are subjected to drifts. Therefore, similar to observatory data, calibrations at regular intervals are required. The method used in space is based on a comparison of the vector data with the absolute scalar data. For this purpose the readings of the vector components $(B_x, B_y, B_z)$ are expanded for all the influencing effects and compared to the total field $|\mathbf{B}|$ derived from the scalar instrument:

$$|\mathbf{B}|^2 = \left[f_x B_x - B_{ox} - B_y cos(x,y) - B_z cos(x,z)\right]^2 \qquad (2.1)$$
$$+ \left[f_y B_y - B_{oy} - B_z cos(y,z)\right]^2 + \left[f_z B_z - b_{oz}\right]^2,$$

where $f_x$, $f_y$, $f_z$ are the scale factors, $B_{ox}$, $B_{oy}$, $B_{oz}$ are the offsets of the components, and $(x,y)$, $(x,z)$, $(y,z)$ represent the angles between the respective sensor elements. Ideally, all these angles should be 90°. Based on a set of measurements, typically extending over one day, the nine fluxgate parameters (3 scale factors, 3 offsets, 3 angles) can be derived from Equation 2.1 in an inversion analysis. This well-established method, called scalar calibration, is routinely used in all magnetic field missions and is described in more detail by Olsen et al. (2003), for example. After applying the new 9 parameters to the fluxgate readings we have a set of absolutely calibrated vector field data.

The next step is to find the orientation of the measured field vector. For this purpose the readings of the star camera are used. Images of the sky are taken, and by comparison of the observed star field with the Hipparcos star catalog (ESA, 1997) the viewing direction is deduced. Here again the star catalog is considered as a well maintained standard. Finally, the vector data are transformed from the celestial coordinates to an Earth-oriented system, such as the NEC (North-East-Center) frame. A detailed description of the transformation matrices including all the elements of Earth rotation is provided and maintained by the International Earth Rotation Service (IERS) (McCarthy, 1996). Applying the transformation to the necessary precision requires the knowledge of the measurement time to a few milliseconds and the position to a few meters. Both these quantities are derived from the GPS receiver.

### 2.2.4 Comparison of Ground and Satellite Data

A data set corrected in the way as described earlier is a good basis for geomagnetic field modeling. Here we shortly introduce the characteristics of magnetic field measurements obtained from ground or from space. Ideally, the data set used in a global analysis has to fulfil the following requirements:

1. Samples should be taken at evenly distributed points, dense enough to cover the shortest wavelengths.
2. Readings at different locations should be taken simultaneously.
3. Measurements should be made in areas free of electric currents, i.e. free of sources of induced magnetic fields.

In principle, all three conditions could be satisfied by a dense ground-based network. In reality, the observatory network is by far not dense enough and rather unevenly distributed. The weakness of (1) can be cured by satellite measurements. Employing magnetic field observations from space has, however, also its shortcomings. For example, requirements (2) and (3) are not obeyed. Satellites have to do the measurements sequentially orbit-by-orbit, and it takes several days or even months to achieve an appropriate global coverage. In order to compensate for this weakness, temporal changes of the magnetic field during the considered interval have to be corrected or co-estimated. Violating requirement (3) is particularly severe since satellites flying in the ionosphere are surrounded by various kinds of currents. For the mitigation of this problem two procedures are followed, one is to remove the magnetic effects of these currents from the recorded data and the second is to select time intervals for field modelling when the currents, which cannot be corrected, are sufficiently weak. In Sect. 2.4 we come back to the characteristics of ionospheric and magnetospheric currents in more detail.

Another limitation inherent to surveys from satellites is that the measurements have to be performed at a certain altitude. This implies that magnetic features smaller than the orbit height cannot be resolved well because the spatial decay of magnetic fields with distance from the source depends on spatial wavelength. Considering recent satellites, the typical altitudes are on average about 800 km for Ørsted, 700 for SAC-C, and 400 km for most of the CHAMP mission, but coming down to 300 km during the final mission phase. To demonstrate the impact of orbit altitude on different wavelength features, attenuation factors are listed in Table 2.1 for SH degrees 20, 50, and 80, relative to observations at the Earth's surface. From this table it becomes clear how important measurements from low orbits are for high-degree magnetic field models. Present improvements in the lithospheric mag-netization models are therefore based entirely on CHAMP observations.

**Table 2.1** Attenuation factors for small-scale crustal features at different orbital altitudes relative to amplitudes at the Earth's surface

| Altitude | deg20 | deg50 | deg80 |
|---|---|---|---|
| Surface | 1 | 1 | 1 |
| 300 km | 0.36 | $9.1 \times 10^{-2}$ | $2.3 \times 10^{-2}$ |
| 400 km | 0.26 | $4.2 \times 10^{-2}$ | $6.8 \times 10^{-3}$ |
| 700 km | 0.10 | $4.4 \times 10^{-3}$ | $1.9 \times 10^{-4}$ |
| 800 km | 0.07 | $2.1 \times 10^{-3}$ | $6.1 \times 10^{-5}$ |

Magnetic field measurements obtained by various platforms have their strengths and weaknesses. By properly combining the data, a maximum of information about the characteristics of the geomagnetic field can be retrieved.

## 2.3 A Potpourri of Magnetic Field Models

Geomagnetic field models offer a global description of the magnetic field based on observational data from ground and/or satellites. They are a useful tool for many practical and scientific purposes, for example, predicting the field distribu-tion at locations or times without direct observations or studying the field evolution and its underlying processes deep within the Earth. Traditionally, magnetic field models describe the geomagnetic core field (e.g. IGRF), while some of the latest models include parts of the long-wavelength lithospheric field and descriptions of contributions from sources external to the Earth (e.g. CM4, the POMME series, CHAOS). Global lithospheric field models (e.g. the MF series) from satellite data are also improving rapidly with the growing amount of satellite data. In this section we present some recent models and what has been learnt from them. A list of websites with digital resources like model coefficients, further descriptions, and software for using the presented models is given in the appendix.

## 2.3.1 Field Representation by Spherical Harmonics and the IGRF

In 1839, Gauss developed spherical harmonic analysis, which is still the most commonly used method to derive global geomagnetic field models. A detailed description of this technique is given, for example, by Langel (1987). In a source-free region the magnetic field $\mathbf{B}$ can be written as the negative gradient of a scalar potential $V$,

$$\mathbf{B} = -\nabla V, \tag{2.2}$$

which satisfies Laplace's equation

$$\nabla^2 V = 0. \tag{2.3}$$

For the field originating inside the Earth, the potential $V_i$ can be developed in a series of spherical harmonic functions:

$$V_i(r, \theta, \phi) = \sum_{n=1}^{n_{max}} \sum_{m=0}^{n} R_E \left( \frac{R_E}{r} \right)^{n+1} [g_n^m \cos m\phi + h_n^m \sin m\phi] P_n^m(\theta), \tag{2.4}$$

where $R_E$ is the Earth's radius, $r, \theta, \phi$ are the spherical coordinates radius, colatitude, and longitude, and $P_n^m$ are the Schmidt quasi-normalized associated Legendre functions of degree $n$ and order $m$. Core and lithospheric field models are parameterized by sets of Gauss coefficients, $g$ and $h$. They can be determined from the data, for example, by a least squares inversion. If the data were measured over a certain time interval there are two commonly used methods to take into account the secular variation. The Gauss coefficients can be expanded in a Taylor series in time, $t$, about the epoch $t_0$ as

$$g(t) = g + (t - t_0)\dot{g} + 0.5(t - t_0)^2 \ddot{g}..., \tag{2.5}$$

and the same can be done for $h(t)$. Thus, coefficients for secular variation, $\dot{g}, \dot{h}$, and secular acceleration , $\ddot{g}, \ddot{h}$, can be obtained. Alternatively, the Gauss coefficients can be expressed as functions of time, for example, by polynomials or cubic B-splines (Bloxham and Jackson, 1992). In cases of sparse data coverage, in time-dependent models and for the higher degrees of secular variation and acceleration sometimes a regularization to constrain smooth behavior is applied, in order to suppress artificial structure and influences from noise.

Fields originating outside the sphere of observation can likewise be expressed by a potential $V_e$:

$$V_e(r,\beta,\lambda) = \sum_{n=1}^{n_{max}} \sum_{m=0}^{n} R_E \left(\frac{r}{R_E}\right)^n [q_n^m \cos m\lambda + s_n^m \sin m\lambda] P_n^m(\beta), \quad (2.6)$$

for which the coefficients generally are named $q$ and $s$ (Langel, 1987). Here, it is advisable to chose a different frame, more suitable to describe the external sources, for the spherical coordinates $(r,\beta,\lambda)$. Again, temporal variations can be taken into account by a suitable temporal description of the coefficients.

Mayer and Maier (2006) presented a different approach of modeling the Earth's magnetic field, using vector kernel functions, which are called vector scaling functions and wavelets. The major ingredient of this representation is a system of vector spherical harmonics, which separates a given spherical vector field with respect to its sources, that is, the spherical vector field is separated into a part which is generated by sources inside the reference sphere, a part which is generated by sources outside the sphere, and a part which is generated by sources on the sphere, which are, for example, currents crossing the sphere. Using this special system of vector spherical harmonics vector scaling functions and wavelets can be constructed, which keep the advantageous property of separating with respect to source but which also allow a locally adapted modeling of the respective vector field.

The most widely known geomagnetic field model, which, for example, is commonly used to determine declination values for navigation by a magnetic compass, is the IGRF, the International Geomagnetic Reference Field. It is based on using a classical spherical harmonic analysis, and it includes the main field and linear secular variation since 1900. A new IGRF model is adopted based on several candidate models every five years by a working group of the International Association of Geomagnetism and Aeronomy (IAGA). The latest version is the IGRF 10th generation, with a new main field model for 2005 and a secular variation estimate to extrapolate up to 2010 (IAGA Working Group V-MOD, 2005). The IGRF is very useful as a standard reference model, but for most scientific applications there exist models that are better tailored for specific purposes, as shown in the following.

## 2.3.2 Satellite Data-Based Field Models

### 2.3.2.1 From Magsat Models to POMME

Magsat was the first satellite to provide geomagnetic vector data covering the whole globe in 1979/1980. These data improved the resolution of global main field models from spherical harmonic degree 10 to 13 and provided secular variation estimates to approximately degree 8 (Langel and Estes, 1985b). With increasing amounts of satellite data, provided by Ørsted, CHAMP and SAC-C, and improved corrections for external field influences, main field and secular variation models have been continuously improving.

Satellite data, like any geomagnetic measurements, include contributions from all internal and external field sources. To derive good main field descriptions, the external field influences have to be eliminated or taken into account in the modeling procedure. It is impossible to separate the contributions in individual measurements. More details about our current understanding of the different external fields are given in Sect. 2.4. The basic method to eliminate external field influences is to use only magnetically quiet night-time data selected on the basis of magnetic activity indices.

The Ørsted Initial Field Model (OIFM) (Olsen et al., 2000) was developed that way for epoch 2000.0, and claimed to be reliable up to spherical harmonic degree 14. The first high resolution model including a secular variation description is the OSVM model for epoch 2000.0 by Olsen (2002), which also includes a first attempt to model the large scale magnetospheric field contributions. This and all following models describe the main field down to the low degree crustal field, and it resolves linear secular variation robustly to spherical harmonic degree 11. Using 14 months of the Ørsted mission and applying classical selection criteria based on local time and external magnetic activity, Langlais et al. (2003) computed a 29-degree internal field model and a 13-degree secular variation model for the period 1999–2000. The CO2 model for epoch 2001.0 (CHAMP, Ørsted, Ørsted2/SAC-C) (Holme et al., 2003) took advantage of the vastly increased amount of data from all three satellites and also provided secular variation to spherical harmonic degree 13.

It soon became clear, however, that the elimination of influences from magnetospheric current systems in all models was insufficient. The following series of POMME (POtsdam Magnetic Model of the Earth) models was specifically addressing the proper treatment of large-scale external fields by a new approach of modeling external fields in different, more suitable coordinate systems. Maus et al. (2005) showed that Solar-Magnetic (SM) coordinates are most appropriate to parameterize the ring current effect and that the outer magnetospheric contributions are best described in the Geocentric-Solar-Magnetospheric (GSM) frame. In the second generation of POMME models the magnetospheric field description was further improved (Maus and Lühr, 2005). Moreover, effects of ionospheric currents on the night side were considered (see Sect. 2.4) and ocean tidal signals were removed (Kuvshinov and Olsen, 2005). The latest model, POMME 3 (Maus et al., 2006), is exclusively based on CHAMP magnetic field data covering the period August 2000 to August 2005, centred on epoch 2003.0 with reliable secular variation and secular acceleration estimates up to spherical harmonic degrees 14 and 10, respectively.

Figure 2.4 shows this improvement in terms of spherical harmonic resolution from the different models. Secular variation as well as acceleration spectra are better clearly resolved up to higher degrees in the recent models. Details of the time-dependent CHAOS model, for which spectra are also included, are given in the following section.

(a)

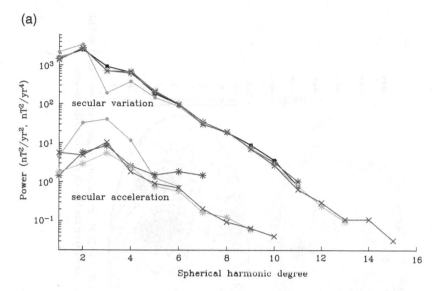

(b)

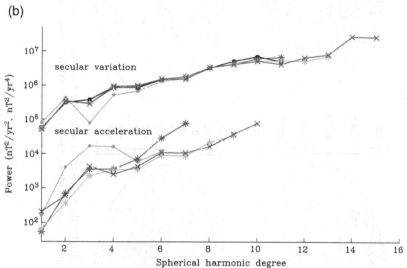

**Fig. 2.4** Spectra of secular variation and acceleration (**a**) at the Earth's surface and (**b**) at the core-mantle boundary (CMB). Models from Magsat data (*gray*), OSVM (*black*), POMME 1.4 (*blue*), POMME 3 (*green*), and CHAOS (for 2002.5, *red*), see text. Only nonregularized coefficients are shown for OSVM and both POMME versions, but due to the time-dependent character of CHAOS, a temporal regularization constraining secular acceleration is present there

### 2.3.2.2 Current Core Field and Secular Variation Characteristics

Major applications of these models are the removal of certain field contributions from the satellite data for studying certain processes, such as electric current systems, ocean flows, or magnetic pulsations based on the remaining signals. The

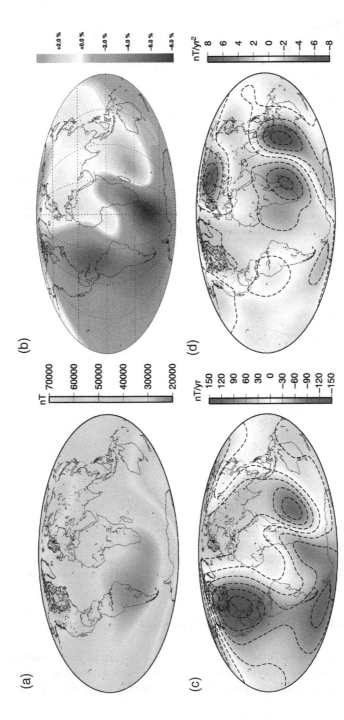

**Fig. 2.5** (**a**) Geomagnetic field intensity 2003.0, (**b**) percentage change of field intensity between 1980 and 2003, (**c**) intensity secular variation 2003.0, and (**d**) intensity secular acceleration 2003.0. Panels (**a**), (**c**) and (**d**) are based on the POMME 3 model and panel (**b**) on the Magsat and POMME 3 models

models also provide a detailed snapshot of the present internal field distribution and its secular change, shown in Fig. 2.5. The field (Fig. 2.5a) is dominated by the structure of a tilted dipole, with maximum field strength in the polar regions and dropping to about half at the equator. In detail, however, significant deviations are observed. In the northern hemisphere there are two maxima rather than just one and a deep depression exists in field intensity in the South America–South Atlantic region. In this so-called South Atlantic Anomaly (SAA) the field exhibits only 60% of the dipole strength at comparable latitudes. This local deficit in field intensity allows energetic particles and cosmic rays to penetrate deeper into magnetosphere and atmosphere than in other regions. Significant space weather effects like satellite outages are frequently observed there (Heirtzler et al., 2002).

Globally, the magnetic field is currently decreasing significantly (see Sect. 2.3.3). However, the secular change is distributed quite unevenly over the globe. Both the snapshot of current secular variation (Fig. 2.5c) and the percentage change between two global field models we computed from Magsat and CHAMP data (Fig. 2.5b) show that over the Eurasian continents and the Indian Ocean the field strength changes little and/or even increases. Strong intensity decrease occurs over North America and in the South Atlantic region. Weaker decrease is observed over the Pacific. The plotted 23-year difference can only represent average changes over this time interval, but secular variation itself is variable. The snapshot of secular acceleration for 2003.0 (Fig. 2.5d) shows that the strongest change of secular variation is currently taking place over the Indian Ocean and Northern Asia. Some changes in the global secular variation pattern are expected to occur over decadal and longer times, and suitable representations to study these are time-dependent models.

## 2.3.3 Investigating Secular Variation: Temporally Continuous Models

### 2.3.3.1 Models gufm, $C^3FM$, CHAOS, CM4, CALS7K

The development of time-dependent models to study details of magnetic core field changes was pioneered by Bloxham and Jackson (1992), which led to a continuous model of the core field for the time span 1590 to 1990 (gufm1) (Jackson et al., 2000), based on modern and historical ground, shipboard, and satellite field measurements. This model is currently being improved and updated to recent epochs (Finlay et al., 2006). A dedicated secular variation model of unprecedented spatiotemporal resolution on the decadal time-scale has been developed within the DFG Priority Programme by Wardinski and Holme (2006). This model, named $C^3FM$ (Continuous Covariant Constrained-end-points Field Model), covers the time interval from 1980 to 2000; at the endpoints it is constrained by models based on satellite data with maximal global data coverage. The observatory and repeat station data available for the years in between cannot provide such global data density, but offer the temporal coverage important for resolving secular variation on short time-scales. Care has been taken to avoid external field influences in this model by considering error covariances between different field elements at each location. Due to the physically motivated regularization involved in gufm1 and $C^3FM$, these two models are particularly suitable for downward continuation to CMB , a handy property of spherical harmonic analysis, which allows the study of the geodynamo processes at their source. The first continuous model with full satellite data coverage is CHAOS (the name referring to the three satellites CHAMP, Ørsted, and SAC-C), covering the time-span March 1999 to December 2005 (Olsen et al., 2006). This model employs state of the art techniques for treatment of satellite data and reliably resolves recent secular variation up to spherical harmonic degree 16 (see Fig. 2.4). It is particularly well-suited for studies of the shortest time characteristics of the geodynamo.

Efforts to model not only the core field, but the complete quiet-time magnetic field time-dependently, including the (static) lithospheric field, magnetospheric and ionospheric fields, and their induced counterparts and influences from coupling currents between ionosphere and magnetosphere, were begun by Sabaka and Baldwin (1993). The latest version is CM4 (Comprehensive Model, Phase 4), covering the time span 1960 to mid-2002 based on all available ground and satellite data (Sabaka et al., 2004).

In an endeavor to bridge the gap between studying recent and paleomagnetic secular variation and to extend global secular variation descriptions to longer time-spans the techniques used to model the recent core field have also been applied to archaeo- and paleomagnetic data of the past millennia. Starting with a series of snapshot models for the past 3000 years (Constable et al., 2000) these efforts led to the development of a continuous model named CALS7K (Continuous model of Archaeomagnetic and Lake Sediment data for the past 7ka), covering the time span from 5000BC to 1950AD (Korte and Constable, 2005). Due to dating uncertainties and high measurement errors inherent in these kind of data, compared to direct field observations, the reliable resolution of this model is spatially limited to about spherical harmonic degree 4, and its temporal resolution is of the order of 100 years.

### 2.3.3.2 Geomagnetic Jerks

All the continuous models have significantly improved our understanding of geomagnetic secular variation on various time-scales, including the shortest-term features known as geomagnetic jerks. They are characterized by sharp changes of secular variation slopes (maxima or minima) within one to two years (Fig. 2.6), and were firstly discovered in the eastward component of European observatory data from around 1969 (Courtillot et al., 1978). Seven such events have been reported for the past 100 years (Alexandrescu et al., 1995; Macmillan, 1996; Mandea et al., 2000) based on geomagnetic observatory data. Several studies have suggested that these impulses are global phenomena (e.g. Malin and Hodder, 1982). However,

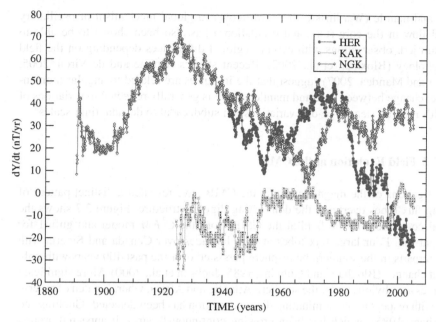

**Fig. 2.6** Time series of secular variation (first derivative of the field) show several sharp changes, known as geomagnetic jerks. Several of the known jerks are most clearly seen in the eastward component (Y) at European locations, like Niemegk observatory (NGK, *blue*) in Germany. Time series from other regions like Hermanus observatory (HER, *red*) in South Africa or Kakioka observatory (KAK, *green*) in Japan show different characteristics with a few clear jerks visible at different times

at least some of the jerks seemed to show a certain time delay between different regions, particularly between the northern and southern hemisphere (Alexandrescu et al., 1996), and are noticeable to different degree in the magnetic vector components. The $C^3FM$ model was designed particularly to study the short-time secular variation. This model suggests more secular variation than previously thought to occur on sub-decadal scales and it shows further jerk-like features, although less prominent, at different locations and times (Wardinski, 2004). Moreover, recently it was discovered that a jerk occurred in 2003, well-covered by magnetic satellite data (Olsen and Mandea, 2007). The CHAOS model has been used for the first global study of a jerk based on a truly dense data coverage of the whole Earth. Similarly to several of the features seen in the $C^3FM$ model, this impulse, seen most clearly in the vertical component, clearly appears to be a regional phenomenon.

The origin of geomagnetic jerks is still unclear. Suggested regional mechanisms include sudden acceleration of the fluid motion at the CMB (Huy et al., 2000), instabilities at the CMB caused by density heterogeneities in the topmost layer of the core (Bellanger et al., 2001), and sudden flux expulsion at the CMB resulting in magnetic diffusion and currents in the lower-most mantle, leading to sudden acceleration or deceleration of the core, due to angular momentum exchange between core and mantle (Wardinski, 2004). Moreover, a global mechanism of oscillatory zonal flow in the core (torsional oscillations) has also been shown to be able to explain jerk observations with certain regional differences depending on the field morphology (Bloxham et al., 2002). Recent studies (Holme and de Viron, 2005; Olsen and Mandea, 2007) suggest that the impulses are related to angular momentum exchange between core and mantle, which is generally reflected in variations of Earth's rotation (length-of-day variations) on subdecadal to decadal time-scales.

### 2.3.3.3 Field Evolution at the CMB

Investigations of the magnetic flux at the CMB have revealed a distinct pattern of strong flux lobes overlying the dominating dipolar structure. Figure 2.7 shows the radial component of the field at the CMB from the $C^3FM$ model and gufm1 for comparison. Four large flux lobes at high latitudes over Canada and Siberia with counterparts in the southern hemisphere persisted over the past 400 years with only slight changes (Bloxham and Gubbins, 1985; Jackson et al., 2000). More significant changes are obvious under the southern Atlantic and Africa, where a patch of reverse flux with respect to the dominating dipole direction has been detected (Gubbins and Bloxham, 1985), which has been growing continuously since it appeared around 1965 (Jackson et al., 2000). Its detailed behavior was investigated by Wardinski and Holme (2006), who found that this patch has enlarged significantly by first combining with a smaller patch below Antarctica. In 1997, this reverse flux patch connected to the northern hemisphere main area of the same flux sign. Another reverse flux patch seen under the Arctic split in two in 1986, with both patches increasing in strength.

### 2.3.3.4 Decrease of the Dipole Moment

The large reverse flux patch in the southern hemisphere plays an important role in the currently observed strong decrease of the dipole field (Gubbins, 1987; Hulot et al., 2002). Since systematic global measurements were begun about 175 years ago, the dipole moment has decreased at a rate of about 5% per century. This rate is about five times faster than that expected for free decay of the dipole if the geodynamo ceased operation, which has lead to some concerns that we might be observing the early stages of a geomagnetic field reversal (Olson, 2002). However, the CALS7K model indicates a high variability of the dipole moment on centennial to millennial time-scales (Korte and Constable, 2006) and recent work by Gubbins et al. (2006) confirmed a significantly lower dipole decay rate prior to 1840. Both this and paleomagnetic considerations suggest that a reversal is not truly imminent (Constable and Korte, 2006). Nevertheless, a substantial further decrease of dipole moment may be a likely future scenario with important consequences for space weather influences on Earth.

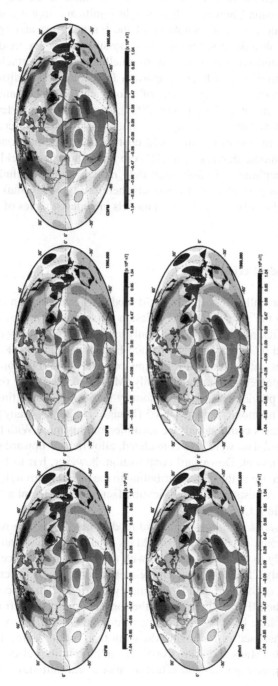

**Fig. 2.7** Radial component of the geomagnetic field at the CMB for 1985 (*left*), 1990 (*center*), and 1995 (*right*) from the $C^3FM$ (*top*) and gufm1 models (*bottom*, present model ends 1990)

## 2.3.4 Lithospheric Field: The Details Are Sharpening

So far we have focused on the characteristics and evolution of the magnetic core field. The satellite missions, however, also brought significant improvements in retrieving the large-scale magnetic signatures of the Earth's crust and upper mantle. Due to the flight altitude of satellites it is impossible to resolve field structure of less than a few hundred kilometers from satellite data. Nevertheless, high-degree spherical harmonic models are important for a number of different purposes, like interpretation and verification of geological provinces and tectonic boundaries (see e.g. Ravat and Purucker, 2003), large wavelength levelling of aeromagnetic surveys and marine survey track lines and reduction of crustal magnetic field signatures in studies of other field contributions. As mentioned earlier, around spherical harmonic degrees 13 to 15 core and lithospheric field contributions have similar amplitudes. Lithospheric field models should include as little core field influence as possible. Inverse models therefore typically start at spherical harmonic degree 14 to 16, while forward models include estimates of the longer wavelengths.

### 2.3.4.1 The MF Model Series

Initial high-degree, so-called lithospheric field models, were based on POGO and Magsat data. Due to the short MAGSAT mission duration and limitations of the scalar data from POGO, reliable expansions only up to SH degrees between 45 and 60 could be achieved (Cain et al., 1989; Arkani-Hamed et al., 1994; Ravat et al., 1995). Real advances came along with the low-orbit CHAMP mission. The first version of the high-degree (MF) model family, MF1, based on only one year of CHAMP scalar data provided a magnetic anomaly map valid at an altitude of 438 km and for spherical harmonic degrees 14–65 (Maus et al., 2002). Many prominent crustal magnetic features could already be identified on this map. With the second version the vector field data were also considered, allowing for upward/downward continuations of the model. Downward continuation, however, has to be regarded with caution, as any noise in the data contributing to the model is strongly amplified by this process. The corresponding amplification factors for different altitudes can be inferred from Table 2.1.

The latest in the series of lithospheric field models is MF5, which is expanded from spherical harmonic degree 16 to 100 (Maus et al., 2007) and can be downward continued nearly to the Earth's surface with reasonable confidence. This was achieved by using only CHAMP data from the most recent 3 years, when the satellite was in its lowest orbit so far, and by constantly improved data processing over the development of the MF model series. All the contributions coming from sources other than the lithosphere have to be eliminated from the CHAMP recordings. The POMME 3 model was used for removing the main field, up to degree 15, and the magnetospheric fields (Maus and Lühr, 2005). In addition to the selection of quiet time data, the polar electrojet far-field effect at sub-auroral latitudes (Maus et al.,

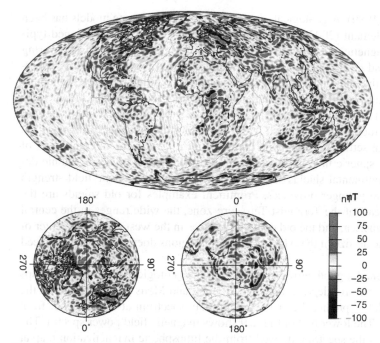

**Fig. 2.8** Vertical component (positive downward) of the large-scale lithospheric field at an altitude of 50 km above the geoid, derived from the MF5 model. The lower two plots show views on the North Pole (*left*) and South Pole (*right*) down to 45° latitude

2005) and magnetic signatures due to ocean tides (Tyler et al., 2003) are corrected for. Although the main part of the ring current effect had already been reduced by the POMME 3 model, there are still significant biases present between neighboring tracks. These are accounted for by a dedicated line levelling procedure. This track-by-track baseline adjustment unfortunately attenuates the signal power, but it brings out individual features significantly clearer. Figure 2.8 shows the large-scale lithospheric field at an altitude of 50 km above geoid height. Prominent crustal magnetic anomalies, which are easily identified on this map are, among others, the Bangui anomaly, the iron ore deposits at Kursk, Kiruna and Kentucky, as well as the clear contrast between the relatively featureless ocean basins and the old, magnetized continental shields.

### 2.3.4.2 Analysis of the Global Lithospheric Field

Methods to obtain geological interpretations of global lithospheric field models include the use of a priori information in inverse models of the lithosphere or forward modelling of the estimated effect of geological structures with comparison and adjustment to inverse models obtained from satellite data. For the Magsat models, global interpretation studies were for example carried out by Meyer et al. (1983) and

Purucker et al. (1989). A geological interpretation of the MF-type models has been performed by Hemant (2003) and Hemant and Maus (2005). They assigned typical values of magnetic susceptibility to known regional geological bodies, covering all continents and ocean plateaus. Induced magnetization was estimated by applying a representative main field to these bodies and the resulting magnetic signatures at satellite altitude could then be compared to those observed in the MF models. Good agreement between the field distributions predicted by the forward modelling and by the MF model were obtained with only minor adjustments in shape of the initial geological settings. In the upper part of Fig. 2.9, the vertical component of the derived lithospheric field is shown. Without truncation of higher harmonic degrees the old continental shields stick out quite clearly in magnetic field strength in comparison to younger provinces. Prominent examples for old shields are the part of Europe east of the Tornquist-Teisseyre zone, the wide ranges in the central part of North America, and the old craton provinces in the west and in the center of Australia. The scale size of these highly magnetic regions does not generally exceed 3,000 km.

Having obtained a good correlation between the geological and magnetic maps in the spectral range for degrees 16 to 80, Hemant and Maus (2005) have used the geological model to predict the magnetic anomaly spectrum at the low end, from degrees 1 to 15. The lower part of Fig. 2.9 shows magnetic field power spectra. The red curve reflects the spectrum derived from the lithospheric magnetization map at satellite altitude. For spherical degrees beyond $n = 16$, it tracks quite closely the observed spectrum (in green, MF2 model). Earlier, lithospheric spectral estimates that include the long wavelengths and are based on a number of different approaches mostly predict a strong drop-off in power in the long-wavelength part (degree 1 to 15) towards global scales, but the details vary significantly (e.g. Cohen and Achache, 1990; Arkani-Hamed et al., 1994; Ravat et al., 1995; Walker and Backus, 1997; O'Brien et al., 1999; Lowe et al., 2001; Korte et al., 2002). The spectral shape obtained by Hemant and Maus (2005) is in good agreement with a theoretical prediction by Jackson (1994), which is based on considerations of correlation length scales.

The high-resolution lithospheric field model from CHAMP data is, moreover, a vital ingredient in the international efforts, sponsored by IUGG (International Union of Geodesy and Geophysics), to produce a World Digital Magnetic Anomaly Map (WDMAM) . This map is based on all available magnetic anomaly data from airborne, marine, ground, and satellite measurements. Surveys at or near the Earth's surface constrain the short spatial wavelengths, while global satellite data are necessary to constrain the long wavelengths properly and enable the merging of regional scale anomaly compilations. Several groups have developed candidates for WDMAM, the GAMMA model proposed by GFZ being one of them (Hamoudi et al., 2007). Based on the candidates, the first WDMAM has just been published in 2007 (Korhonen et al., 2007). This WDMAM represents the total field magnetic anomalies of the Earth's lithosphere at a grid interval of 3 minutes of latitude/longitude and altitude of 5 km above the geoid, with wavelengths corresponding to spherical harmonic degree 16 to 7200.

(a)

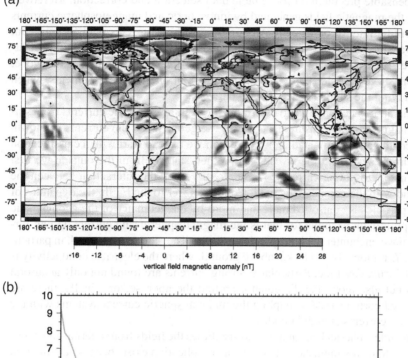

(b)

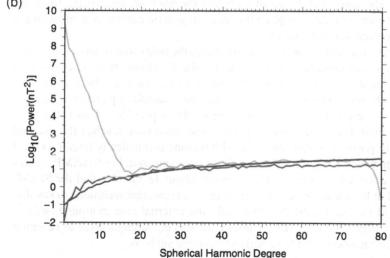

**Fig. 2.9** (a) Magnetic anomalies of the vertical component at satellite altitude predicted by forward modelling of induced magnetization based on susceptibility estimates for geological structures. (b) power spectra of the model in (a) (*red*), compared to MF2 with core field (*green*) and a theoretical model (*blue*, Jackson, 1994)

## 2.4 Characterization of External Field Sources

To determine the fields having their sources in the core or lithosphere, contributions from electric currents flowing above the Earth's surface are regarded as disturbances. A proper understanding of the magnetic fields from external sources is an indispensable precondition for reliable data selection and correction. Moreover, studying the magnetic fields generated by ionospheric and magnetospheric currents is of scientific interest on its own. Among others, these currents play an important role in transferring energy and momentum from the solar wind into the upper atmosphere. They furthermore help to distribute the solar wind input over the globe. The recent survey missions provided the high resolution field measurements, which allow studying the subtle details of these currents.

In general, it is not possible to determine the feature of electric currents from single site magnetic field measurements, but if the geometry of the currents is known, their strength can be estimated. For that reason we have to start with a classification of the current systems to be investigated. When starting from the outside we first find the magnetospheric currents. These are driven by the interaction between the solar wind with the geomagnetic field. Coming closer to the Earth, the dynamics of charged particles is controlled more and more by the geometry of the geomagnetic field. A prominent example for that is the magnetospheric ring current. Going further down we encounter the ionospheric currents . They are concentrated, in particular, in the $E$ region, about 110 km above ground, where the electrical conductivity is maximal during day-time. Particularly, strong currents are found not only at auroral latitudes but also along the dip-equator around the noon sector. Finally, there are field-aligned currents (FAC) coupling the magnetospheric current systems with the ionospheric currents at high latitudes.

In principle, spherical harmonic analysis allows the fields from internal and external sources to be distinguished. In reality, a complication exists because the internal and external fields are expressed in different reference frames, which move with respect to each other. Therefore, very different data sampling patterns in the two frames have to be considered. Polar orbiting satellites provide a reasonable sampling of the main field within one day. Over the same time interval the external fields are only poorly sampled, since the orbital plane is practically fixed in a local time frame, in which most of the external fields are expressed. For CHAMP it takes 130 days to cover all local times, Ørsted needs about 18 months, and the SAC-C orbit is fixed in local time. When applying the field separation technique to satellite magnetic field measurements, therefore, only the external contributions from stable average currents are accessible. Currents of varying strength have to be either parameterized time-dependently or omitted from the analysis.

## 2.4.1 Magnetospheric Fields

Prominent sources of magnetic disturbances at the Earth's surface are magnetospheric currents. A commonly used indicator for the ring current strength is the $D_{ST}$ (storm-time disturbance) index (for its definition, see Sugiura, 1965). Accurate modeling of the internal field from near-Earth magnetic measurements requires considering this contribution properly. Based on Magsat data, Langel and Estes (1985a) deduced a linear relation of the large-scale external field with the $D_{ST}$ index.

Quantitatively, they attributed 63% of the $D_{ST}$ value to an external source and 27% of this external signal to the induction effect. In addition, they identified a constant part of about 19 nT adding to the $D_{ST}$ dependent part. These first degree external and internal fields were found to be well aligned with the geomagnetic dipole axis. Due to the sun-synchronous orbit of Magsat (dawn/dusk) and the short mission, duration studies of any local time, seasonal, or solar cycle dependencies of the external field signature could not be performed. Consequently, this simple magnetospheric field approximation, a symmetric ring current varying in strength linearly with $D_{ST}$, has been in use until 2000 (e.g. the Ørsted Initial Field Model, Olsen et al., 2000). More detailed models of the external field distribution have been developed based on the large amount of high-precision geomagnetic measurements from satellites such as Ørsted and CHAMP.

Olsen (2002) introduced in his modeling approach annual and semi-annual terms not depending on $D_{ST}$. The origin of these fields was not fully clear at that time. Significant progress in characterizing the external contributions came from modeling the different fields in their appropriate coordinate systems. By applying this technique when developing the POMME 2 model (cf. Sect. 2.3.2) Maus and Lühr (2005) found that the near-Earth effect from magnetospheric currents can conveniently be separated into two parts, one of which is handled in GSM (Geocentric Solar Magnetospheric) and the other in SM (Solar Magnetic) coordinates. In the inner magnetosphere the ring current is the most prominent system. Since the shape of this current is closely controlled by the geomagnetic field, it can quite efficiently be described in the SM frame. Dominant current systems in the distant magnetosphere (e.g. magnetospheric tail) are driven primarily by the interaction of the solar wind with the geomagnetic field. Their effect is therefore better described in the GSM system. In this frame the Earth performs a wobbling motion comprising both a diurnal and annual component. Figure 2.10 provides a schematic view of the main magnetospheric current systems contributing to the near-Earth magnetic field. The contribution of the tail currents during nonstorm times is well represented by a stable field of 13 nT pointing southward, aligned with the GSM $z$-axis. The magnetic contribution in GSM was attributed to currents in the magnetospheric tail. It amounts to 13 nT during quiet times and is well aligned with the GSM $z$-axis, pointing southward. The IMF magnetic field $B_y$ component additionally causes a small tilt of the field vector in the $y - z$ plane of about $1°/nT$. At a given observatory the GSM field causes both diurnal and annual variation with amplitudes of the order of 5 nT. In studies of ionospheric currents, such as the $S_q$ (solar quiet) current system, the diurnal contributions of the GSM field should be removed before the magnetic signatures are interpreted. The annual variation of the GSM field has been observed before in quiet night-time observatory data (e.g. Sillanpää et al., 2004) and has earlier been termed "seasonal baseline variation" (Campbell, 1989). This contribution should be subtracted from observatory measurements before using them in modeling efforts.

The other part of the magnetospheric field, best treated in SM coordinates, is attributed to the combined effect of ring current and day-side magnetopause currents. It consists of a stable part causing a field close to the Earth of about 8 nT, which is

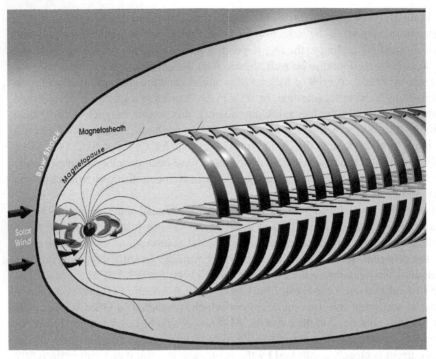

**Fig. 2.10** Schematic illustration of major magnetospheric current systems contributing to the near-Earth magnetic field. At a distance of about 5 Earth radii the westward directed ring current is circling the Earth. On the sunward side, Chapman-Ferraro currents are flowing on the magnetopause. The current system on the night side consists of cross-tail currents flowing from the dawn to dusk side, and closing by magnetopause currents over the northern and southern tail lobes

approximately aligned with the dipole axis and is pointing southward. The variable part of the SM fields is found to be proportional to the $D_{ST}$ index. The much larger ring current field bias reported by Langel and Estes (1985a) reflects probably the sum of stable SM and GSM contributions.

These external field variations induce secondary currents in the Earth's crust and mantle. Maus and Weidelt (2004) extended the simple approach of Langel and Estes (1985a) and proposed a splitting of the $D_{ST}$ value into its external and internal part, termed $E_{ST}$ and $I_{ST}$ index, respectively (see Appendix). With the help of these two parameters the global distribution of the ring current effect can be computed more precisely. The induced part from the SM field was further treated by Olsen et al. (2005b) who proposed a frequency dependent transfer function between inducing and induced part. This procedure removes in particular the constant part from the new $I_{ST}$ index. There are, however, still uncertainties remaining, such as the reliability of the conductivity model on which the transfer function is based. We cannot tell what the real constant value of $I_{ST}$ is because the $D_{ST}$ values have been calculated for only about 50 years.

## 2.4.2 Ionospheric Currents at Mid-Latitudes

There are well-known current systems in the ionosphere. The major constituents are shown in Fig. 2.11. On the dayside at mid-latitudes, these are the $S_q$ and the equatorial electrojet (EEJ) currents (for reviews see e.g. Campbell (1989) and Forbes (1981), respectively). These currents are confined to the ionospheric $E$-layer. The characteristics of the currents have been reproduced in several electrodynamics models. In our context, their magnetic effects at the Earth's surface and in space are of more interest. A rather reliable representation of the average magnetic field from the $S_q$ and EEJ currents is provided by the CM4 model (Sabaka et al., 2004). In CM4 the $S_q$ and EEJ currents are treated as a coupled system, which, however, does not in general seem to be the case (Manoj et al., 2006). A more detailed investigation of the EEJ was presented by Lühr et al. (2004). Although the $S_q$ currents are a daytime phenomenon, they can partly also influence the magnetic field measurements on the night side. Due to the diurnal variation of the $S_q$ current, mirror currents are induced in the crust and upper mantle. These induced currents respond with a

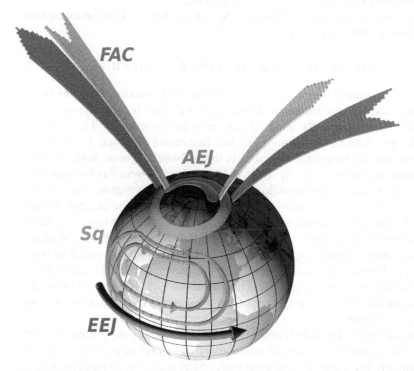

**Fig. 2.11** Schematic illustration of the main ionospheric current systems. At high latitudes the auroral electrojets (AEJ) flow predominantly from noon to midnight both on the dawn and dusk side. The electrojets are flanked by FAC. At mid-latitudes the $S_q$ currents are flowing on the dayside. A current vortex with opposite sense exists in the southern hemisphere. The equatorial EEJ is a narrow feature of intense eastward currents

phase delay. After sunset, when the ionospheric currents have ceased flowing, the currents in the Earth start to flow in the reverse direction over several hours, just to maintain a zero-mean average over a day. With that they generate a small but persistent magnetic field during the night. A possible field leakage into geomagnetic field models has been investigated by Olsen et al. (2005a). They compared two models, CM4, which takes the $S_q$ and its induction effect into account and OSVM, which is based on Ørsted night side measurements without considering any ionospheric contributions. As expected, a field pattern antisymmetric to the dip equator emerges. The residuals in the radial direction are of the order of 3 nT, pointing upward in the northern and downward in the southern hemisphere. The authors concluded that most present-day geomagnetic field models suffer from this contamination.

Recently, it has been found that there are also sizable currents flowing in the ionospheric $F$ region on the night side (e.g. Lühr et al., 2003; Maus and Lühr, 2006), at the same height range in which the satellites are operating. A careful reduction of the ionospheric effects in satellite data is therefore required to avoid spurious effects in main field models. So far, it was commonly assumed that ionospheric currents at mid-latitudes can be ignored at night time because of the reduction in $E$ layer conductivity by about two orders of magnitude.

There are several effects contributing to the current density, $\mathbf{j}$, in the ionosphere (see e.g., Kelley, 1989, Sect. 4.2)

$$\mathbf{j} = \underline{\sigma}\,(\mathbf{E} + \mathbf{u} \times \mathbf{B}) + [nm_i\mathbf{g} \times \mathbf{B} - k_B\nabla\,((T_i + T_e)n) \times \mathbf{B}]\,\frac{1}{B^2} \qquad (2.7)$$

where $\underline{\sigma}$ is the conductivity tensor, $\mathbf{E}$ is the electric field including both, the large-scale background field and the polarization electric field that is required to maintain the current continuity; $\mathbf{u}$ is the wind velocity, $n$ is the electron density, $m_i$ the ion mass, $\mathbf{g}$ the gravitational acceleration, $k_B$ the Boltzmann constant, $T_e$ and $T_i$ are the electron and ion temperatures, and $\mathbf{B}$ is the ambient magnetic field. At night the transverse conductivity largely disappears and the first term can be neglected. The first term in square brackets reflects the gravity-driven currents. These are not depending on any formal conductivity but vary primarily with the electron density. The first observational evidence for the significance of this current type, sometimes called "ionospheric ring current", was presented by Maus and Lühr (2006). It attains its maximum current density slightly above the F2 layer density peak and at latitudes following the diurnal and seasonal variation of the equatorial ionization anomaly. At CHAMP altitude the magnetic effect amounts to about 5 nT. Figure 2.12 shows the global distribution of the quantities controlling the intensity of the eastward flowing gravity-driven currents.

The second term in brackets is related to currents driven by the plasma pressure gradient. This is also independent of conductivity and therefore present at night. It affects primarily the magnetic field magnitude. After sunset steep electron density gradients build up at the bottom side of the F region. At lower altitudes the recombination of ions is much faster than in the F region, which is practically removing the E-layer. In addition there is the ion fountain effect at low latitudes, which lifts up the ionosphere along the dip equator and forms the Appleton anomalies , ribbons of

IRI electron density and IGRF H/B²: 21.03.2001, 18UT, 400km

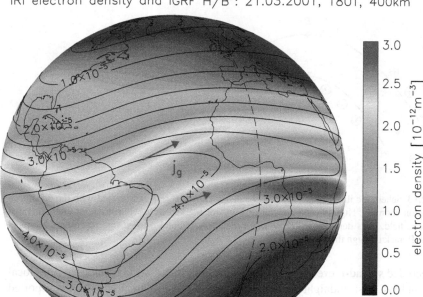

**Fig. 2.12** Global distribution of electron density and the ratio of horizontal component $H$ to the square of the field strength $B$ (in nT), both controlling the gravity-driven current density, $j_g$. The *two arrows*, next to $j_g$, mark the centers and the direction of the gravity-driven currents. The *dashed line* indicates the 18:00 local time meridian

enhanced plasma density on both sides of the equator. This anomaly is particularly well developed during the hours from sunset till midnight. The pressure gradient currents flow in a sense that they reduce the magnetic field strength inside the region of enhanced plasma density. This effect is also termed the diamagnetic effect of dense plasmas. Figure 2.13 depicts schematically the configuration of diamagnetic currents encircling the flux tubes of the equatorial ionization anomaly, and it shows the generated magnetic field. The general importance of pressure gradient currents for precise magnetic field measurements in the ionosphere was shown by Lühr et al. (2003). The authors reported a magnetic field deficit of up to 5 nT in dense plasma regions, and they offered a first-order correction approach.

Currents associated with local plasma irregularities were first identified in CHAMP data by Lühr et al. (2002). In a subsequent study, Stolle et al. (2006) investigated the magnetic signatures of these so-called equatorial spread-F events, also known as plasma bubbles. In particular, they could determine specific climatological features such as the spatial distribution, the dependencies on local time, season, and solar activity. CHAMP crosses these depleted flux tubes typically at latitudes about 10° north and south of the equator. Figure 2.14 shows the global distribution of

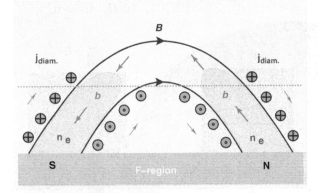

**Fig. 2.13** Schematic illustration of the diamagnetic currents, $j_{\text{diam}}$, associated with the equatorial ionization anomaly. High plasma density regions are marked by $n_e$. Furthermore, the generated magnetic field, $b$, is depicted by *green arrows*. The horizontal *dashed line* represents a typical CHAMP track through the system

all recorded spread-F events. The plasma irregularities start to form after 1900 local time and last past midnight. A distinct seasonal, longitudinal variation is reported by the authors. They find high occurrence rates over Africa during June solstice, which shift to South America around December solstice. During equinox seasons there is an almost even longitudinal distribution. Furthermore, a strong dependence

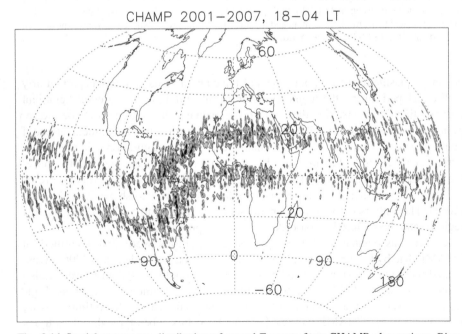

**Fig. 2.14** Spatial occurrence distribution of spread-F events from CHAMP observations. Bins ($1 \times 1°$ size) with warm colors representing high and cool color low occurrence rates

on solar activity is observed. The average occurrence frequency shows a linear dependence on the F10.7 index. Because of their rather distinct magnetic signatures (systematic field enhancement; appearance at confined latitude bands; seasonal longitudinal occurrence variation), these spread-F events can cause spurious effects if not accounted for in magnetic field modeling efforts.

The night side F region currents described earlier significantly disturb main field modeling, not only because of their magnetic fields. Since they flow at satellite altitude, they violate the assumption of measurements in a current-free space. An efficient removal of these currents is required in order to avoid spurious effects in a model that is interpreted in terms of a scalar potential. More dedicated investigations are required to obtain a better description of these ionospheric currents.

### 2.4.3 The High-Latitude Ionosphere During Quiet Times

Rather difficult areas for main and crustal field modeling are the auroral regions. Here intense ionospheric currents are flowing both on the day and night side. As depicted in Fig. 2.11, the electrojets flow along the auroral oval from the dayside to the nightside. Prime drivers for the high latitude current systems are the strong FAC flanking the electrojets. A number of studies have been devoted to characterizing the auroral current systems (e.g. Glassmeier, 1987; Untiedt and Baumjohann, 1993; Paschmann et al., 2003). Here we focus on the ionospheric features during times with weak or vanishing auroral currents. Measurements taken during quiet periods are particularly valuable for field modeling.

It is not a simple task to identify periods of low activity at high altitudes. It has been known for quite a while and was demonstrated by Ritter et al. (2004b) that the commonly used planetary activity index, $Kp$, or the $D_{ST}$ index are not suitable for identifying periods of weak auroral currents. Even the Polar Cap index (PC) is not very efficient. On the other hand, the auroral electrojet intensity, which is responsible for most of the disturbances in data for field modeling, can be estimated equally well from ground and satellite measurements (Ritter et al., 2004a). This opens the opportunity to predict the magnetic effect of this current system at satellite location from ground observations. Unfortunately, the required dense networks of magnetometers are available only in the European and American sectors. In other words, no global estimate of the auroral activity can be provided. For that reason this approach has not been pursued any further.

In a quest for suitable indicators identifying periods in which the auroral region is quietest, we have tested among others the intensity of FAC. Ritter and Lühr (2006) showed that the FAC density derived from CHAMP measurements and the locally observed deviations of the total field strength are well correlated. At high latitudes only the field magnitude is commonly considered in internal magnetic field modelling. The close relationship between FAC strength and total field disturbances suggests that the estimated FAC density might be a suitable activity indicator for high latitudes. Unfortunately, the good correlation vanished when restricting the analysis to samples of low FAC densities. Obviously, FAC flowing further away from the satellite track also contribute to the local electrojet intensity.

Magnetic reconnection at the magnetospause causes a magnetospheric electric field. This merging electric field, $E_m$, was tested as a further indicator as it is a measure for the coupling efficiency between solar wind and magnetosphere. It is defined as (e.g., Kan and Lee, 1979)

$$E_m = v_{sw}\sqrt{B_y^2 + B_z^2}\sin^2(\theta/2) \qquad (2.8)$$

where $v_{sw}$ is the solar wind velocity, $B_y$ and $B_z$ are the IMF components in GSM coordinates, and the clock angle is $\theta = \arctan(B_z/B_y)$. According to merging theory this electric field is mapped to the day-side auroral region and drives currents there. The intensity of the resulting currents should be proportional to the product of conductivity times electric field. Periods of low activity at auroral latitudes are thus expected in darkness, when there is no photo-ionization, and in addition, no reconnection is going on. We have taken a closer look at the magnetic perturbations recorded by CHAMP within the polar night during the 40 days centred on the hemispheric winter solstice. Ritter and Lühr (2006) confirm that the magnetic perturbations stay below 10 nT when $E_m$ is less than 0.8 mV/m. This selection criterion is, however, only efficient in the day time sector between 08 and 16 magnetic local time (MLT). For comparison, Fig. 2.15 shows on the left side the distribution of peak total field (dF) disturbance detected by CHAMP from all passes of the South Pole around local winter. Here, a clear concentration of the disturbances along the auroral oval emerges. On the right side of the figure detected dF deflections are shown for

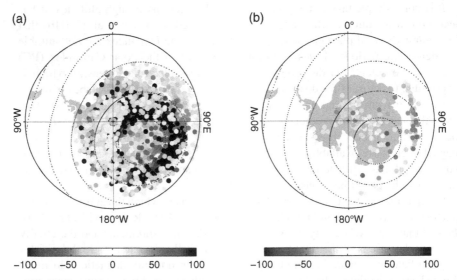

**Fig. 2.15** Distribution of peak total field, dF, disturbance detected by CHAMP (in nT). On the left side, all passes of the South Pole around local winter are shown. On the right side only most undisturbed passes were selected.

passes that are selected by our stringent selection criterion: 40 days around winter solstice, $E_m < 0.8$ mV/m and local time sector 08–16 MLT. Remaining dF peaks are now unrelated to the auroral oval and indicate probably genuine crustal magnetic signals. The selection scheme introduced here is probably too stringent for the determination of the main field with its secular variation since it provides readings from the high latitudes only for two short periods every year alternating between the two hemispheres. For the determination of the lithospheric magnetization distribution, which in general is regarded to be stationary (e.g. Langel and Hinze, 1998), this selection criterion seems appropriate.

## 2.5 Open Issues and Outlook

The early years of this millennium have seen the advent of a new era of space-borne geomagnetic field research. The fleet of spacecraft, Ørsted, CHAMP, and SAC-C returned a wealth of unprecedented high-resolution magnetic field measurements. Thanks to international initiatives like the *Decade of Geopotential Research* and national programmes like the priority programme *Geomagnetic Field Variations* the international geomagnetic community could take advantage of new data, much as the activities of the *Göttinger Magnetischer Verein* stimulated geomagnetic research in the nineteens century. The analysis of these new data changed and improved our view of the recent magnetic field significantly. A number of impressive achievements have been described in this chapter. There are, however, still several open issues, which are waiting to be addressed in the future.

Among the most crucial are those linked to improving the separation of external from internal field contributions. In the case of magnetospheric currents, for example, the magnetic effect of substorms on field modeling has never been investigated. Commonly applied data selection criteria do not exclude these phenomena. Another important issue concerns the F region currents on the night side. For their understanding, a self-consistent model of ionospheric processes is required. With a better understanding of the current drivers it may be possible to develop more suitable correction approaches. Furthermore, little attention has been paid to the large-scale FAC. They are causing toroidal magnetic fields, which cannot be described by scalar potentials. Correcting for that effect is a major task for future missions. The expertise gathered and the results obtained so far are an excellent investment in the preparation for tackling future research objectives in this field.

A challenge immediately ahead of us is the optimal utilization of the final CHAMP mission data, recorded during the low-orbit cruising phase shortly before reentry. According to orbit predictions, these 300-km altitude orbits are expected toward the end of 2008, coinciding with the solar minimum. This fortunate coincidence promises a high yield of unperturbed magnetic field measurements from CHAMP, which will further improve the resolution of crustal magnetization maps.

Despite the achievements on main field and secular variation descriptions, we are still far from understanding in detail the underlying core processes. The increasing, uninterrupted satellite data series will lead to further refined descriptions of detailed short-term secular variation as a basis to study core flow and to test hypotheses about mechanisms causing rapid changes. The future looks bright for gaining a better understanding of the geodynamo, perhaps even allowing for predictions of future decadal field changes. On the one hand, there are the ongoing improvements of global geomagnetic field models, both in detail and toward longer time-scales. On the other hand, present and expected numerical geodynamo simulations get more detailed and show already several features that agree with real observations. The combination of numerical simulations with constraints from actual data models appears promising. Moreover, an increasing number of studies on the magnetic fields of other planets offers opportunities for comparisons and analyses of similarities and differences.

The high-resolution magnetic field monitoring of the geomagnetic field from space will continue even after the end of all the present survey missions. The European Space Agency (ESA) has selected the Swarm satellite mission as its fifth Earth Observation Opportunity Mission. This constellation mission consists of three CHAMP-like spacecraft flying in a dedicated formation and providing simultaneous measurements from three different local times and at two different altitudes. Their launch is planned for 2010 with an expected mission duration of at least 4 years. The period of high-resolution field recording will be extended significantly and Swarm will provide an even better picture of the processes controlling the geodynamo. Another focus of the Swarm mission will be the further refinement of lithospheric magnetic anomaly maps. Two spacecraft, flying side-by-side in a low orbit, will enable new approaches for field modeling. The consideration of the magnetic field gradient, derived from the measurements of two adjacent spacecraft in the analysis, promises highly enhanced resolution. With all the experience gained during the past fruitful years, the Swarm mission can be expected to yield many more significant scientific results.

## Appendix: Digital Resources

Here we provide a list of web-pages containing additional information on several of the models described earlier, including coefficients and in some cases software or online-calculators to obtain field values. A website describing the ongoing international World Digital Magnetic Anomaly Map (WDMAM) project is included.

**IGRF - The International Geomagnetic Reference Field**
http://www.ngdc.noaa.gov/IAGA/vmod/igrf.html
http://www.ngdc.noaa.gov/seg/geomag/models.shtml

**Ørsted satellite data models OIFM, OSVM**
http://spacecenter.dk/research/solar-system-physics/orsted-1/Magnetic-models

**Multisatellite data models CO2, CHAOS**
http://spacecenter.dk/research/solar-system-physics/orsted-1/Magnetic-models
http://www.gfz-potsdam.de/pb2/pb23/Models/model.html

**CHAMP satellite data models POMME**
http://www.gfz-potsdam.de/pb2/pb23/Models/model.html
http://geomag.colorado.edu/model.html
**Hourly values of the $E_{ST}/I_{ST}$ indices**
http://www.ngdc.noaa.gov/seg/geomag/est_ist.shtml).
The indices have also been determined for past times from 1957 onward.
**High resolution secular variation model C3FM**
http://www.gfz-potsdam.de/pb2/pb23/Models/model.html
**Lithospheric field models MF5 and predecessors**
http://www.gfz-potsdam.de/pb2/pb23/Models/model.html
http://geomag.colorado.edu/model.html
**Information on WDMAM**
http://projects.gtk.fi/WDMAM/index.html
**Comprehensive Models CM4 and predecessors**
http://core2.gsfc.nasa.gov/CM/
**Long-term models CALS7K and predecessors**
http://www.gfz-potsdam.de/pb2/pb23/Models/model.html
http://igpphome.ucsd.edu/ cathy/Projects/Holocene/holocene.html

# References

Alexandrescu, M., Gibert, D., Hulot, G., Mouël, J.-L. L., and Saracco, G. (1995). Detection of geomagnetic jerks using wavelet analysis. *J. Geophys. Res.*, 100:12557–12572.

Alexandrescu, M., Gibert, D., Hulot, G., Mouël, J.-L. L., and Saracco, G. (1996). Worldwide wavelet analysis of geomagnetic jerks. *J. Geophys. Res.*, 101:21975–21994.

Arkani-Hamed, J., Langel, R., and Purucker, M. (1994). Scalar magnetic anomaly maps of the Earth derived from POGO and Magsat data. *J. Geophys. Res.*, 99:24,075–24,090.

Bellanger, E., Mouël, J.-L. L., Mandea, M., and Labrosse, S. (2001). Chandler wobble and geomagnetic jerks. *Phys. Earth Planet. Int.*, 124:95–103.

Bloxham, J. and Gubbins, D. (1985). The secular variation of the Earth's magnetic field. *Nature*, 317:777–781.

Bloxham, J. and Jackson, A. (1992). Time-dependent mapping of the magnetic field at the core-mantle boundary. *J. Geophys. Res.*, 97:19,537–19,563.

Bloxham, J., Zatman, S., and Dumberry, M. (2002). The origin of geomagnetic jerks. *Nature*, 420(6911):65–68.

Cain, J., Wang, Z., Kluth, C., and Schmitz, D. (1989). Derivation of a geomagnetic model to n = 63. *Geophys. J. Int.*, 97:431–441.

Campbell, W. (1989). The regular geomagnetic field variations during quiet solar conditions. In Jacobs, J. A., editor, *Geomagnetism*, volume 3, pp. 385–460. Academic Press, Orlando.

Cohen, Y. and Achache, J. (1990). New global vector magnetic anomaly maps derived from Magsat data. *J. Geophys. Res.*, 95:10783–10800.

Constable, C. and Korte, M. (2006). Is Earth's magnetic field reversing? *Earth Planet. Sci. Lett.*, 246:1–16.

Constable, C. G., Johnson, C. L., and Lund, S. P. (2000). Global geomagnetic field models for the past 3000 years: transient or permanent flux lobes? *Phil. Trans. R. Soc. Lond. A*, 358:991–1008.

Courtillot, V., Ducroix, J., and Mouël, J.-L. L. (1978). Sur une accélération récente du la variation séculaire du champ magnétique terrestre. *C.R. Acad. Sci. Paris Ser. D*, 287:1095–1098.

European Space Agency (1997). *The Hipparcos and Tycho Catalogues*. ESA SP-1200.

Finlay, C., Jackson, A., and Gillet, N. (2006). An updated version of the historical field model gufm1. In *10th Symposium on Study of the Earth's Deep Interior*, p. 68.

Forbes, J. (1981). The equatorial electrojet. *Rev. Geophys.*, 19:469–504.

Gauss, C. (1833). *Intensitas vis magneticae terrestris ad mensuram absolutam revocata*. Sumtibus Dieterichianis, Göttingen.

Gauss, C. (1839). Allgemeine Theorie des Erdmagnetismus. In *Resultate aus den Beobachtungen des Magnetischen Verein im Jahre 1838*, pp. 1–52. Göttinger Magnetischer Verein, Leipzig.

Glassmeier, K. H. (1987). Ground-based observations of field-aligned currents in the auroral zone - Methods and results. *Ann. Geophys.*, 5:115–125.

Gubbins, D. (1987). Mechanism for geomagnetic polarity reversals. *Nature*, 326:167–169.

Gubbins, D. and Bloxham, J. (1985). Geomagnetic field analysis - III. Magnetic fields on the core-mantle boundary. *Geophys. J. R. Astron. Soc.*, 80:695–713.

Gubbins, D., Jones, A., and Finlay, C. (2006). Fall in Earth's magnetic field is erratic. *Science*, 312:900–902.

Hamoudi, M., Thebault, E., Lesur, V., and Mandea, M. (2007). GeoForschungsZentrum Anomaly Magnetic MAp (GAMMA): A candidate model for the World Digital Magnetic Anomaly Map. *Geochem., Geophys., Geosys.*, 8, Q06023:doi:10.1029/2007GC001638.

Heirtzler, J., Allen, J., and Wilkinson, D. (2002). Ever-present South Atlantic Anomaly damages spacecraft. *EOS, Trans. AGU*, 83:165.

Hemant, K. (2003). *Modelling and interpretation of global lithospheric magnetic anomalies*. PhD thesis, Freie Univ., Berlin.

Hemant, K. and Maus, S. (2005). Geological modeling of the new CHAMP magnetic anomaly maps using a Geographical Information System (GIS) technique. *J. Geophys. Res.*, B, 110, B12103:doi:10.1029/2005JB003837.

Holme, R. and de Viron, O. (2005). Geomagnetic jerks and a high-resolution lenght-of-day profile for core studies. *Geophys. J. Int.*, 160:435–439.

Holme, R., Olsen, N., Rother, M., and Lühr, H. (2003). CO2: A CHAMP magnetic field model. In Reigber, C., Lühr, H., and Schwintzer, P., editors, *First CHAMP Mission Results for Gravity, Magnetic and Atmospheric Studies*, pp. 220–225. Springer, Berlin - Heidelberg.

Hulot, G., Eymin, C., Langlais, B., Mandea, M., and Olsen, N. (2002). Small-scale structure of the geodynamo inferred from Ørsted and Magsat satellite data. *Nature*, 416:620–623.

Huy, M. L., Mandea, M., Mouël, J.-L. L., and Pais, A. (2000). Time evolution of the fluid flow at the top of the core. Geomagnetic jerks. *Earth, Planets, Space*, 52:163–173.

IAGA Working Group V-MOD (2005). The 10th generation international geomagnetic reference field. *Geophys. J. Int.*, 161:561–656.

Jackson, A. (1994). Statistical treatment of crustal magnetization. *Geophys. J. Int.*, 119:991–998.

Jackson, A., Jonkers, A., and Walker, M. (2000). Four centuries of geomagnetic secular variation from historical records. *Phil. Trans. R. Soc. Lond. A*, 358:957–990.

Jankowski, J. and Sucksdorff, C. (1996). *IAGA guide for magnetic measurements and observatory practice*. IAGA.

Kan, J. and Lee, L. (1979). Energy coupling function and solar wind-magnetosphere dynamo. *Geophys. Res. Lett.*, 6:577–580.

Kelley, M. (1989). *The Earth's Ionosphere*. Elsevier, New York.

Korhonen, J., Fairhead, J., Hamoudi, M., Hemant, K., Lesur, V., Mandea, M., Maus, S., Purucker, M., Ravat, D., Sazonova, T., and Thebault, E. (2007). *Magnetic Anomaly Map of the World*. Map published by Commission for Geological Map of the World supported by UNESCO, 1st Edition.

Korte, M. and Constable, C. (2006). Centennial to millennial geomagnetic variation. *Geophys. J. Int.*, 167:43–52.

Korte, M., Constable, C., and Parker, R. (2002). Revised magnetic power spectrum of the oceanic crust. *J. Geophys. Res.*, 107, B9, 2205:doi:10.1029/2001JB001389.

Korte, M. and Constable, C. G. (2005). Continuous geomagnetic field models for the past 7 millennia: 2. CALS7K. *Geochem., Geophys., Geosys.*, 6, Q02H16:doi:10.1029/2004GC000801.

Kuvshinov, A. and Olsen, N. (2005). 3-D modelling of the magnetic field due to ocean tidal flow. In Reigber, C., Lühr, H., Schwintzer, P., and Wickert, J., editors, *Earth Observation with CHAMP, Results from Three Years in Space*, pp. 359–365. Springer, Berlin - Heidelberg.

Langel, R. (1987). The main field. In Jacobs, J. A., editor, *Geomagnetism*, Vol. 1, pp. 249–512. Academic Press, Orlando.

Langel, R. and Estes, R. (1985a). Large-scale, near-Earth magnetic field from external sources and the corresponding induced internal field. *J. Geophys. Res.*, 90:2487–2494.

Langel, R. and Estes, R. (1985b). The near-Earth magnetic field at 1980 determined from Magsat data. *J. Geophys. Res.*, 90:2495–2509.

Langel, R. and Hinze, W. (1998). *The Magnetic Field of the Earth's Lithoshpere: The Satellite Perspective*. Cambridge Univ. Press.

Langlais, B., Mandea, M., and Ultré-Guérard, P. (2003). High-resolution magnetic field modeling: Application to MAGSAT and Ørsted data. *Phys. Earth Planet. Int.*, 135:77–92.

Lowe, D. A. J., Parker, R. L., Purucker, M. E., and Constable, C. G. (2001). Estimating the crustal power spectrum from vector Magsat data. *J. Geophys. Res.*, 106:8589–8598.

Lühr, H., Maus, S., and Rother, M. (2004). The noon-time equatorial electrojet, its spatial features as determined by the CHAMP satellite. *J. Geophys. Res.*, 109, A01306:doi:10.1029/2002JA009656.

Lühr, H., Maus, S., Rother, M., and Cooke, D. (2002). First in-situ observation of night-time F region currents with the CHAMP satellite. *Geophys. Res. Lett.*, 29(10):doi:10.1029/2001GL013845.

Lühr, H., Rother, M., Maus, S., Mai, W., and Cooke, D. (2003). The diamagnetic effect of the equatorial Appleton anomaly: Its characteristics and impact on geomagnetic field modelling. *Geophys. Res. Lett.*, 30(17):doi:10.1029/2003GL017407.

Macmillan, S. (1996). A geomagnetic jerk for the early 1990's. *Earth Planet. Sci. Lett.*, 137:189–192.

Malin, S. and Hodder, B. (1982). Was the 1970 jerk of internal or external origin? *Nature*, 296:726–728.

Mandea, M., Bellanger, E., and Mouël, J.-L. L. (2000). A geomagnetic jerk for the end of the 20th century? *Earth Planet. Sci. Lett.*, 183:369–373.

Manoj, C., Lühr, H., Maus, S., and Nagarajan, N. (2006). Evidence for short spatial correlation lengths of the noon-time equatorial electrojet – inferred from a comparison of satellite and ground magnetic data. *J. Geophys. Res.*, 111, A11312:doi:10.1029/2006JA011855.

Maus, S. and Lühr, H. (2005). Signature of the quiet-time magnetospheric magnetic field and its electromagnetic induction. *Geophys. J. Int.*, doi:10.1111/j.1365–246X.2005.02691.x.

Maus, S. and Lühr, H. (2006). A gravity-driven electric current in the Earth's ionosphere identified in CHAMP satellite magnetic measurements. *Geophys. Res. Lett.*, 33, L02812:doi:10.1029/2005GL024436.

Maus, S., Lühr, H., Balasis, G., Rother, M., and Mandea, M. (2005). Introducing POMME, the Potsdam Magnetic Model of the Earth. In Reigber, C., Lühr, H., Schwintzer, P., and Wickert, J., editors, *Earth Observation with CHAMP, Results from Three Years in Space*, pp. 293–298. Springer, Berlin - Heidelberg.

Maus, S., Lühr, H., Hemant, K., Balasis, G., Ritter, P., and Stolle, C. (2007). Fifth generation lithospheric magnetic field model from CHAMP satellite measurements. *Geochem. Geophys. Geosys.*, 8, Q05013:doi:10.1029/2006GC001521.

Maus, S., Rother, M., Holme, R., Lühr, H., Olsen, N., and Haak, V. (2002). First scalar magnetic anomaly map from CHAMP satellite data indicates weak lithospheric field. *Geophys. Res. Lett.*, 29(14):doi:10.1029/2001GL013685.

Maus, S., Rother, M., Stolle, C., Mai, W., Choi, S.-C., Lühr, H., Cooke, D., and Roth, C. (2006). Third generation of the Potsdam Magnetic Model of the Earth (POMME). *Geochem. Geophys. Geosyst.*, 7, Q07008:doi:10.1029GC001269.

Maus, S. and Weidelt, P. (2004). Separating the magnetospheric disturbance magnetic field into external and transient internal contributions using a 1D conductivity model of the earth. *Geophys. Res. Lett.*, 31, L12614:doi:10.1029/2004GL020232.

Mayer, C. and Maier, T. (2006). Separating inner and outer Earth's magnetic field from CHAMP satellite measurements by means of vector scaling functions and wavelets. *Geophys. J. Int.*, 167:1188–1203.

McCarthy, D. (1996). IERS conventions. IERS Technical Notes 21, U.S. Naval Observatory.

Meyer, J., Hufen, J., Siebert, M., and Hahn, A. (1983). Investigations of the internal geomagnetic field by means of a global model of the Earth's crust. *J. Geophys.*, 52:71–84.

Neubert, T., Mandea, M., Hulot, G., von Freese, R., Primdahl, F., Jørgensen, J., Friis-Christensen, E., Stauning, P., Olsen, N., and Risbo, T. (2001). Ørsted satellite captures high-precision geomagnetic field data. *EOS, Trans., AGU*, 82:81.

O'Brien, M. S., Parker, R. L., and Constable, C. G. (1999). Magnetic power spectrum of the ocean crust on large scales. *J. Geophys. Res.*, 104:29,189–29,201.

Olsen, N. (2002). A model of the geomagnetic field and its secular variation for epoch 2000 estimated from Ørsted data. *Geophys. J. Int.*, 149:454–462.

Olsen, N., Holme, R., Hulot, G., Sabaka, T., Neubert, T., Tøffner-Clausen, L., Primdahl, F., Jørgensen, J., Léger, J.-M., Barraclough, D., Bloxham, J., Cain, J., Constable, C., Golovkov, V., Jackson, A., Kotzé, P., Langlais, B., Macmillan, S., Mandea, M., Merayo, J., Newitt, L., Purucker, M., Risbo, T., Stampe, M., Thomson, A., and Voorhies, C. (2000). Ørsted Initial Field Model. *Geophys. Res. Lett.*, 27(22):3607–3610.

Olsen, N., Lowes, F., and Sabaka, T. (2005a). Ionospheric and induced field leakage in geomagnetic field models, and derivation of candidate models for DGRF 1995 and DGRF 2000. *Earth Planet Space*, 57:1191–1196.

Olsen, N., Lühr, H., Sabaka, T., Mandea, M., Rother, M., Tøffner-Clausen, L., and Choi, S. (2006). CHAOS – A model of the Earth's magnetic field derived from CHAMP, Ørsted, and SAC-C magnetic satellite data. *Geophys. J. Int.*, 166(1):67–75.

Olsen, N. and Mandea, M. (2007). Investigation of a secular variation impulse in satellite data: The 2003 geomagnetic jerk. *Earth Planet. Sci. Lett.*, 255:94–105.

Olsen, N., Sabaka, T., and Lowes, F. (2005b). New parameterisation of external and induced fields in geomagnetic field modelling, and a candidate model for IGRF 2005. *Earth Planet Space*, 57:1141–1149.

Olsen, N., Tøffner-Clausen, L., Sabaka, T., P. Brauer, J. M., Jørgensen, J., Léger, J.-M., Nielsen, O., Primdahl, F., and Risbo, T. (2003). Calibration of the Ørsted vector magnetometer. *Earth, Plants, Space*, 55:11–18.

Olson, P. (2002). The disappearing dipole. *Nature*, 416:591–594.

Paschmann, G., Haarland, S., and Treumann, R., editors (2003). *Auroral Plasma Physics, Space Sci. Rev.*, 103/1–4.

Purucker, M., Langel, R., Rajaram, M., and Raymond, C. (1989). Global magnetization models with a priori information. *J. Geophys. Res.*, 103:2563–2584.

Ravat, D., Langel, R. A., Purucker, M., Arkani-Hamed, J., and Alsdorf, D. E. (1995). Global vector and scalar Magsat magnetic anomaly data. *J. Geophys. Res.*, 100:20,111–20,136.

Ravat, D. and Purucker, M. (2003). Unraveling the magnetic mystery of the earth's lithosphere. In Reigber, C., Lühr, H., and Schwintzer, P., editors, *First CHAMP Mission Results for Gravity, Magnetic and Atmospheric Studies*, pp. 251–260, Berlin - Heidelberg. Springer.

Reigber, C., Lühr, H., and Schwintzer, P. (2002). CHAMP mission status. *Adv. Space Res.*, 30(2):129–134.

Ritter, P. and Lühr, H. (2006). Search for magnetically quite CHAMP polar passes and the characteristics of ionospheric currents during the dark season. *Ann. Geophys.*, 24:2997–3009.

Ritter, P., Lühr, H., Maus, S., and Villianen, A. (2004a). High-latitude ionospheric currents during very quiet times: Their characteristics and predictability. *Ann. Geophys.*, 22:2001–2014.

Ritter, P., Lühr, H., Viljanen, A., Amm, O., Pulkkinen, A., and Sillanpää (2004b). Ionospheric currents estimated simultaneously from CHAMP satellite and IMAGE ground-based magnetic field measurements: A statistical study at auroral latitudes. *Ann. Geophys.*, 22:417–430.

Sabaka, T. and Baldwin, R. (1993). Modeling the Sq magnetic field from POGO and Magsat satellite and contemporaneous hourly observatory data: Phase I. Contract report HSTX 9302, Hughes STX Corp. for NASA/GSFC Contract NAS5-31 760.

Sabaka, T. J., Olsen, N., and Purucker, M. E. (2004). Extending comprehensive models of the Earth's magnetic field with Ørsted and CHAMP data. *Geophys. J. Int.*, 159:521–547.

Sillanpää, I., Lühr, H., Villianen, A., and Ritter, P. (2004). Quiet-time magnetic variations at high latitude observatories. *Earth Planets Space*, 56:47–65.

Stolle, C., Lühr, H., Rother, M., and Balasis, G. (2006). Magnetic signatures of equatorial spread F, as observed by the CHAMP satellite. *J. Geophys. Res.*, 111, A02304:doi:10.1029/2005JA011184.

Sugiura, M. (1965). Hourly values of equatorial $D_{ST}$ for the IGY. *Ann. Int. Geophys. Year*, 35:9–45.

Tyler, R., Maus, S., and Lühr, H. (2003). Satellite observations of magnetic fields due to ocean tidal flow. *Science*, 299:239–241.

Untiedt, J. and Baumjohann, B. (1993). Study of polar current systems using the IMS Scandinavian magnetometer array. *Space Sci. Rev.*, 63:245–390.

Walker, A. and Backus, G. (1997). A six-parameter statistical model of the earth's magnetic field. *Geophys. J. Int.*, 130:693–700.

Wardinski, I. (2004). *Core surface flow models from decadal and subdecadal secular variation of the main geomagnetic field*. PhD thesis, Free University Berlin.

Wardinski, I. and Holme, R. (2006). A time-dependent model of the Earth's magnetic field and its secular variation for the period 1980–2000. *J. Geophys. Res.*, 111, B12101:doi:10.1029/2006JB004401.

# Chapter 3
# Records of Paleomagnetic Field Variations

Karl Fabian and Roman Leonhardt

## 3.1 Origin and Carriers of Paleomagnetic Signals

The Earth's internal magnetic field is the only direct active signal from Earth's core, which can be detected nearly undistorted at Earth's surface, and it has been perpetually recorded by magnetic remanence carriers in rocks throughout the Earth's history. This rock magnetic record of the paleomagnetic field is an extremely valuable archive. On one hand, it is used for plate-tectonic reconstruction, and provides exact magnetostratigraphic information about motion and chronology of the crust. On the other hand, rock magnetic data constitute the only direct recording of the history of magnetic field generation in the deep interior of the Earth.

Experimental determinations of the paleomagnetic field are now challenged to produce data sets that can be compared to numerical geodynamo models. Only when such data are sufficiently reliable and detailed, it can be decided which of the different imaginable processes within the Earth's outer core most closely resembles reality. It is likely that during the next decades, the fundamental physical mechanisms, which are driving the enormous internal machinery of our planet, will be revealed by the combined effort of the community of geoscientists. The main physical database on which the success of this endeavor depends are rock magnetic determinations of paleofield direction and intensity.

While the determination of the paleofield direction from the stepwise demagnetization of a rock usually can be accomplished without complications, the determination of paleofield intensity is extremely challenging. Unfortunately, even a global cover by directional data would allow only to infer the geometry of the paleofield. In order to assess the activity of the geodynamo, the knowledge of the past field intensity is indispensable.

In consequence, the optimal determination of the paleointensity of the Earth's magnetic field by means of rock magnetic measurements is a central experimental

Karl Fabian
Geological Survey of Norway Leiv Eiriksons vei 39 7491 Trondheim Norwegen

Roman Leonhardt
Department Angewandte Geowissenschaften und Geophysik Lehrstuhl Technische Ökosystemanalyse Universität Leoben Peter Tunner Straße 25-27 8700 Leoben Austria

K.-H. Glaßmeier et al. (eds.), *Geomagnetic Field Variations*, Advances in Geophysical and Environmental Mechanics and Mathematics,
© Springer-Verlag Berlin Heidelberg 2009

problem, which recedes the progress of retracing the history of the geodynamo, and therefore of revealing the physical processes in the Earth's liquid core. Therefore, paleointensity determination lies at the heart of the two main themes of this chapter: (1) to understand the physical origin of the rock magnetic signal in sediments and lava flows, and (2) to relate it to the past variations of the Earth's magnetic field.

The mechanisms of remanence acquisition, and our current state of knowledge about them, are completely different for sediments and igneous rocks.

### 3.1.1 Sedimentary Remanence

In sediments the remanence acquisition occurs primarily by mechanical rotation of already magnetized minerals. Figure 3.1 gives a brief sketch of the formation of DRM (detrital remanent magnetization) and pDRM (post-depositional remanent magnetization). A classical DRM is acquired by magnetized particles, which align to Earth's field while settling through the water column. This occurs very quickly, even when Brownian motion is taken into account (Stacey, 1972). Usually this DRM does not survive in the mixed layer of the sediment. Almost all sedimentary NRM must be considered to be a pDRM. In marine environments, the above single-grain

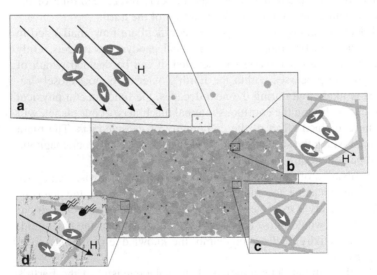

**Fig. 3.1** Schematic model of sedimentary remanence acquisition. Several mechanical processes contribute to the remanence acquisition of sediments. (**a**) Depositional remanence is acquired while magnetic particles settle in the water column. (**b**) During sediment compaction magnetic particles may move freely inside sediment pores, i.e., created by clay minerals, and thereby acquire a post-depositional remanence. The efficiency of this remanence acquisition process depends upon the intensity of mechanical activation. (**c**) Magnetic particle fixed by surrounding clay minerals. (**d**) Bioturbation can lead to realignment of magnetic particles. Varying intensity of bioturbation can change efficiency of remanence acquisition without influencing the matrix composition

sedimentation is also unrealistic, because the small magnetic particles are likely to reach the sediment–water interface in bigger flocculated or coagulated compounds within which the particles are already mechanically or electrostatically bound. The relative remanence directions of the magnetic particles within such a compound are probably uncorrelated and the couple upon the compound due to the total remanence is negligible in comparison to the mechanical couple due to irregular shape or turbulent flow (van Vreumingen, 1993a,b; Katari and Bloxham, 2001). It is unknown how the magnetic particles are fixed and whether they still can acquire a pDRM, although the latter is likely.

In a Bingham model of pDRM acquisition, single magnetic particles are fixed in a visco-elastic non-newtonian (thixotropic) matrix. Shear stress of magnetic origin generates an elastic response until a certain limit stress $\sigma_b$ is reached. At this point a viscous response sets in and the particle rotates toward the magnetic field (Fig. 3.1c,d) until the stress falls again below $\sigma_b$ (Shcherbakov and Shcherbakova, 1987). Results of Tucker (1980, 1981a,b), Otofuji and Sasajima (1981), and others indicate that mechanical activation of the sediment is of major importance for the pDRM acquisition, and subsequently for the NRM intensity. This can be interpreted when a magnetic particle is considered, which is fixed to a matrix surface in the vicinity of a pore space by van-der-Waals forces. By some activation process like bioturbation, Brownian motion (temperature), pore-water flow (compaction), seismicity or some other mechanism, the particle is set free inside the pore space and according to Stacey (1972) aligns quickly within Earth's field until it is fixed again at the pore boundary (Fig. 3.1b) . This model supplements the Bingham model of Shcherbakov and Shcherbakova (1987), and perhaps can be regarded formally within this frame by taking into account a distribution of limit stresses $\sigma_b$ instead of a single value. Moreover, an activation function must be added to represent the above mechanical activation mechanisms.

Independent of its origin, sedimentary sequences often provide densely spaced and accurately dated records of natural remanent magnetization (NRM). Whenever the alternating field (AF) demagnetization of the NRM indicates that a single stable direction of the paleofield has been recorded throughout the sequence, a relative record of the Earth's magnetic field intensity variation can be constructed. This is done by dividing the NRM at depth $z$ by a normalization parameter (normalizer). The normalizer $v(z)$ is chosen such as to be proportional to the concentration of the remanence carriers at depth $z$. Common choices are anhysteretic remanent magnetization (ARM), isothermal remanent magnetization (IRM), or magnetic volume susceptibility ($\chi$). Remanence parameters (ARM and IRM) are generally demagnetized by the same AF field as the NRM (Levi and Banerjee, 1976). The normalization procedure intends to remove the concentration dependence of the NRM and the remaining NRM variations are believed to mirror the paleofield intensity. Therefore, the obtained normalized NRM record is denoted as relative paleointensity (RPI) record.

The practical technique of measurement and interpretation of RPI records has been developed in a series of articles starting with Johnson et al. (1948) and continued among others in Harrison (1966), Harrison and Somayajulu (1966),

Johnson et al. (1975), Levi and Banerjee (1976), and King et al. (1983). A detailed review of the development of RPI techniques can be found in Tauxe (1993), and a recent review of relative paleointensity records has been presented by Valet and Herrero-Bervera (2003).

### 3.1.2 Thermoremanence

All generally accepted methods of absolute paleointensity determination in igneous rocks are based on Thelliers method (Thellier, 1941). The most commonly used version of Coe (1967a) proceeds from the assumption that the rocks NRM is a thermoremanent magnetization (TRM), acquired in ancient times by rapid cooling from the Curie temperature to the ambient temperature under the influence of the external geomagnetic field. This initial TRM is removed in several heating steps to increasing temperatures $T_i$ in zero external field leaving a residual remanence tNRM($T_i$). After each heating step to $T_i$, the sample is again heated to the same temperature $T_i$ but cooled in an exactly known external field $H_{lab}$. Thereby, an additional partial thermoremanence pTRM($T_i$) is acquired. In successful measurements, plotting tNRM($T_i$) against pTRM($T_i$) yields a straight line with slope—$H_{paleo}/H_{lab}$, where $H_{paleo}$ is the required paleointensity.

As a result of extensive experimental studies, Thellier formulated the following three laws, which he assumed to be essential for his method to work.

1. *Linearity of TRM*
   Any TRM or partial TRM (pTRM) is proportional to the field within which it is acquired.
2. *Independence law of pTRM*
   A pTRM acquired between temperatures $T_1$ and $T_2 > T_1$ is not influenced by temperature changes below $T_1$ and is completely removed by any heating above $T_2$.
3. *Additivity law of pTRM*
   The sum of two or more pTRMs acquired separately in nonoverlapping temperature intervals is equal to a single pTRM acquired in the union of these temperature intervals.

Néel gave a satisfying theoretical proof of the above laws for rocks and baked clays provided that remanence is carried by independent SD particles (Néel, 1949). An SD particle has a clearly defined thermal blocking temperature $T_B$ above which during cooling its magnetic moment is unstable and below which it is completely fixed. Moreover, during heating the magnetic moment becomes unstable at the same unblocking temperature $T_{UB} = T_B$.

For PSD or MD remanence carriers abundant in natural rocks, there exists no generally accepted theoretical foundation for absolute paleointensity determination. The lack of such an MD TRM theory has the effect that no standardized rules for evaluation and verification of the Thellier-experiment or at least for preselection of the samples are available. Accordingly, absolute paleointensity determination—due to its small success rate—still is one of the most time consuming and difficult measurements in laboratory geophysics. In essence, the lack of a reliable physical theory of thermoremanence beyond Néel's SD theory is the main reason that the current database of available paleointensity values contains only some 3,000 data points, many of which are incompatible. As a result, even the general outline of the long-term history of geomagnetic field strength is subject to intense scientific dispute.

The remaining chapter reviews collaborative studies on several themes of current paleomagnetic research into the history of Earth's magnetic field variations. The first section covers investigations into the foundations and techniques of paleointensity determination. The second section reports on recent studies of long-term field stability. It presents lava records of intensity variations during the Cretaceous and Neogene and an approach to disentangle the paleointensity signal from environmental influence within a set of Late Quaternary sedimentary records from the Central South Atlantic. Short-term paleomagnetic variations are considered in the last section of this chapter, where the focus lies on reconstructing the field geometry during geomagnetic reversals and excursions.

## 3.2 Techniques of Paleointensity Determination

Paleointensity values of the past Earth magnetic field are usually much more difficult to extract than obtaining directions of that field. The basic reasons for this difficulty are related to a number of conditions, which need to be satisfied during absolute intensity determinations. Basically, two different types of paleointensity determinations are distinguished: (a) absolute paleointensity determinations which allow for retrieving the "true" value of the past geomagnetic field strength and (b) relative paleointensity determinations which allow for obtaining a relative measure of intensity changes within a sampling succession.

### 3.2.1 Absolute Paleointensities

All absolute intensity determination techniques rely on a comparison of an artificial thermoremanence acquired in a known laboratory field with the NRM of the sample. The Thellier method (Thellier and Thellier, 1959) and, in particular, derived modifications (e.g., Coe, 1967b; Aitken et al., 1988) are the most commonly used paleomagnetic techniques for the determination of the intensity of the past Earth's magnetic field.

The Thellier-type techniques rely on demagnetization of NRM plus acquisition of pTRM in a constant laboratory field at stepwise increasing temperatures. Among other techniques, the Shaw method (Shaw, 1974) is the most widespread, which compares the alternating field demagnetization spectrum of NRM and pTRM. Only recently, a multispecimen protocol was suggested for paleointensity determination, based on the measurement of pTRM of multiple samples acquired at a single temperature step at different laboratory field strengths (Dekkers and Boehnel, 2006).

In order to retrieve reliable information from a certain rock, in particular when applying Thellier-type techniques, seven assumptions have to be implied, which are outlined below. These are as follows:

(1) The remanent magnetization of the rock has to be a preserved thermoremanent magnetization acquired in a constant magnetic field with negligible secondary components, hereinafter abbreviated as nTRM.
(2) Partial thermoremanent magnetization (pTRM) acquired during the Thellier experiment, hereinafter abbreviated as pTRM*, are independent and additive.

These basic conditions were first nominated by Thellier (1941) and are commonly referred to as Thellier's laws. The law of additivity states that a pTRM* acquired in a temperature range of $(T_1,T_3)$, further referred to as pTRM*$(T_1,T_3)$, equals the sum of pTRM*$(T_1,T_2)$ and pTRM*$(T_2,T_3)$ with $T_3 > T_2 > T_1$ and with laboratory fields applied only during cooling within the given temperature range. The law of independence requires that a pTRM*$(T_1,T_2)$ is completely removed when heating to $T_2$ in zero field and is not affected by heating and cooling below $T_1$. These laws, however, are only fulfilled for magnetic grains showing single domain (SD) character (Néel, 1949). The failure of Thellier's laws during paleointensity determinations can lead to a concave curvature of the Arai diagram (Coe, 1967a; Levi, 1977; Dunlop and Özdemir, 2000), as well as to s-shaped curves, which show a concave trend in the lower temperature range and a convex trend in the higher temperature range in the Arai diagram (Coe, 1967a; Coe et al., 2004b; Leonhardt et al., 2004b). A concave curvature is linked to a contribution of remanences with $T_{ub} < T_b$, so-called low temperature tails (Dunlop and Özdemir, 2000; Fabian, 2001). Convex curvature is related to high temperature tails (Coe et al., 2004b; Leonhardt et al., 2004b). Furthermore, the failure of Thellier's laws also affects alteration checks, also often referred to as pTRM checks. Despite the absence of alteration, these checks show deviations from the expected value for PSD and MD-like behavior (Leonhardt et al., 2004b). For experiments where the laboratory field is applied parallel to the NRM ($\triangle\theta = 0°$), the cumulative deviation of these pTRM checks is plotted versus the domain state proxy in Fig. 3.2.

These measurements were conducted on thermally stabilized natural basalts and synthetic magnetite samples. The domain-state proxy is calculated from the average deviation to an ideal paleointensity determination.

Failures of Thellier's laws can be observed during the paleointensity experiment using two techniques, the pTRM*-tail check (Riisager and Riisager, 2001), which monitors failures of independency of pTRM*s and the additivity checks (after Krása et al., 2003).

(3) Both, nTRM and pTRM*, are linearly dependent on the applied magnetic field.

It has been shown experimentally by various laboratories that TRM acquisition, independent of grain size, is linearly related to the magnetic field when applying only

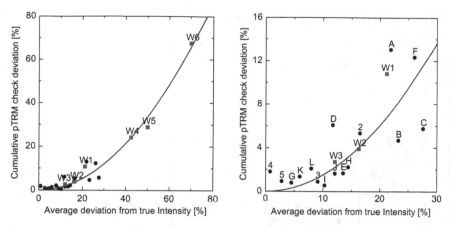

**Fig. 3.2** Cumulative pTRM check deviation normalized to the TRM versus the average deviation from the ideal paleointensity value for the 0° experiments. pTRM checks were measured in-field. In addition to the stabilized samples, results from pure magnetite samples with well-defined grain-size are shown. *Letters and numbers* are sample identifiers; *black symbols* represent thermally stabilized basaltic samples and *red symbols* correspond to Wright magnetic samples from Krása et al. (2003)

weak fields ($< 100 - 200$ $\mu$T). Slight deviations from linearity in weak magnetic field were also reported (Coe, 1967a; Dunlop, 1973). This deviation from linearity, however, is not large enough to affect the intensity determination (Coe, 1967a). When applying laboratory magnetic fields of the same order as the natural ambient field even strong deviations from linearity do not affect the intensity determination.

(4) nTRM and pTRM* are independent of the cooling rate or are acquired with the same cooling rate.

It has been shown theoretically (Dodson and McClelland-Brown, 1980; Halgedahl and Fuller, 1980) and experimentally (Fox and Aitken, 1980; McClelland-Brown, 1984) that an assembly of identical, noninteracting SD particles acquires a larger TRM after slower cooling. Negative interactions and/or MD kinetics, however, lead to a lower TRM after slower cooling rates (McClelland-Brown, 1984). In Fig. 3.3 the magnetic cooling rate dependency for thermally stabilized natural basalts and artificial magnetite samples is shown in relation to the domain state proxy of Fig. 3.2.

For the majority of the samples an increasing domain-state proxy leads to a decreasing ratio of slowly cooled TRM to fast cooled TRM. One MD-like basaltic sample and the largest artificial magnetite sample, however, show strong deviations from this trend, leading to significantly larger TRMs after slow cooling. The reason for this is yet unclear. An unbiased paleointensity determination requires that the cooling rate in the laboratory is chosen similar to the natural cooling rate. As this is not always practical, corrections can be applied if the natural cooling rate is either known or can be determined, for example, using geospeedometry (Leonhardt et al., 2006). Mostly, however, the cooling rate differences between laboratory and nature are neglected.

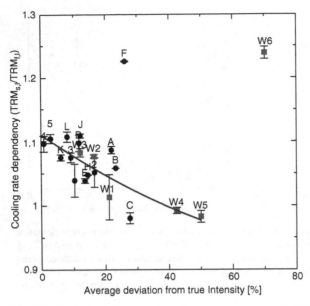

**Fig. 3.3** Magnetic cooling rate dependency versus average deviation from true intensity. The magnetic cooling rate dependency was determined by the ratio of TRM intensities acquired during slow laboratory cooling ($TRM_{s,l}$) and fast laboratory cooling ($TRM_{f,l}$). For testing the thermal stability, the cooling rate experiment was repeated two times and thus enables an error estimation on the $y$-axis. *Letters* and *numbers* are sample identifiers; *black symbols* represent thermally stabilized basaltic samples and *red symbols* correspond to Wright magnetic samples from Krása et al. (2003)

(5) **Magnetic particle interaction affecting the natural and artificial magnetization processes differently must be absent.**

According to a theoretical study of Coe (1974) magnetic interactions do not affect Thellier-type paleointensity determinations if TRM and induced magnetizations are statistically stable, isotropic, and linear with field. These conclusions are supported by the phenomenological model of Fabian (2001). An important assumption for above theoretical approaches and the drawn conclusions is that magnetic particle interaction is independent from magnetic phases and their blocking temperatures. It was shown recently (Krása et al., 2005) that this is not generally true when dealing with multiphase basaltic rocks. Such magnetic interactions lead to a peculiar effect in the NRM/pTRM diagrams. In Fig. 3.4 two paleointensity determinations on the same thermally stabilized sample are shown.

For Fig. 3.4a the laboratory field is applied parallel to the NRM ($\triangle\theta = 0°$). An antiparallel configuration was used for Fig. 3.4b. Both experiments show a decrease of pTRM above the 500°C and a TRM smaller than the pTRM at 500°C. Magnetomineralogical changes are absent and the occurrence of this effect at two different angles between NRM and laboratory field direction also proves that tails are not causing the decreased TRM capacity by approaching $T_C$. This effect occurs only in samples with a major $T_C$ slightly below 500°C. High coercivity proportions

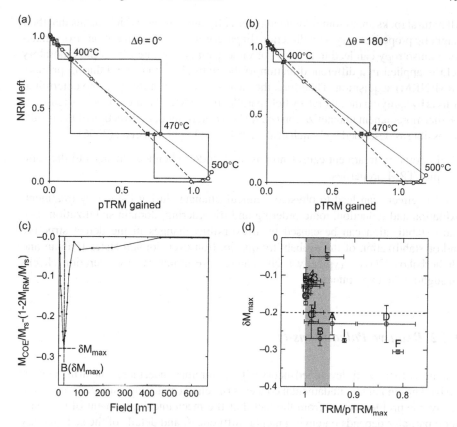

**Fig. 3.4** Paleointensity determinations and rock magnetic parameters for sample D. Axes are normalized to the same intensity for all Arai diagrams. The determinations for $\triangle\theta = 0°$ (**a**) and $180°$ (**b**) both show a decrease of pTRM above 500°C. The parameter $\delta M_{max}$ of the Henkel plot (**c**) is sensitive for interactions. In (**d**) the average difference between TRM and maximal pTRM of steps 3 and 4 is plotted versus the average $\delta M_{max}$ (steps 2,6). Error bars indicate the standard deviation

are virtually absent in those samples ($S_{300} > 0.97$). The most plausible cause for such decreasing pTRM-TRM capacity is an acquisition of pTRM opposite to the applied field direction at temperatures above 500°C, and thus a negative coupling of magnetic particles. This interpretation is further supported by IRM/backfield analyses (Henkel, 1964; Leonhardt, 2006), which has also been suggested to be useful for preselection of suitable paleointensity samples (Wehland et al., 2005) (Fig. 3.4c,d). Interpretation of the highT segment of the paleointensity determination of such samples will lead to an underestimate of paleointensity of more than 30%.

(6) Magnetic field deflection/reductions caused by intrinsic magnetic anisotropy or the sample-demagnetizing field due to its surrounding must have a negligible effect on nTRM and pTRM*.

All natural rocks are to some extend magnetically anisotropic, which means that their magnetic properties vary with direction. In paleointensity determination strong magnetic anisotropy can lead to an over- or underestimate of intensity if the laboratory field is applied in a different direction to the natural ambient field during primary TRM(NRM) acquisition. Therefore, the easiest way to minimize bias to determinations is by applying the laboratory field parallel to the NRM. If the anisotropy tensor of the thermoremanent magnetization (ATRM) is determined, a correction of the results is possible as described by Veitch et al. (1984) and Leonhardt et al. (2006).

(7)  Laboratory treatment causes no physical and/or chemical changes to the samples TRM properties.

Various effects can lead to physicochemical changes during laboratory treatment: oxidation and reduction, ionic ordering and disordering, domain stabilization. Domain stabilization can be caused by irreversible changes in the defect structure and/or stabilization of the probability distribution over domain states (Fabian and Shcherbakov, 2004). Typically such changes are monitored by alteration checks throughout the experiments.

## 3.2.2 Relative Paleointensity

In contrast to the well-developed theory of thermoremanence, there is no firm theoretical background of the fundamental concepts of relative paleointensity determination. Partly, the problem arises from the fact that the mechanical alignment of the magnetic particles depends on environmental influences, and details of the sedimentary matrix. Here we present a linear model, which takes such influences into account and at least qualitatively shows how relative paleointensity determination works.

### 3.2.2.1  A Linear Model of Sedimentary Sequences

Sediment composition varies with depth. To quantify this variation in a linear mixing model, the sediment is described as a collection of $n$ disjoint sedimentary components $P_i$ with $i = 1, \ldots, n$. While concentrations $c_i(z) \geq 0$ of the components $P_i$ change with depth $z$, the physical and especially magnetic properties of the components are assumed to be independent of $z$. The $c_i(z)$ are collected in a concentration vector $\mathbf{c}(z) \in \mathbb{R}^n$, where $\sum c_i \leq 1$. For example, a component $P_i$ may comprise a certain grain-size fraction of a magnetic mineral, e.g., all single-domain magnetite particles.

Yet, it can also represent a complicated compound of matrix and magnetic minerals, which has a well-defined remanence-acquisition behavior, for example, slightly interacting 1 $\mu$m TM10 particles in foraminiferal ooze with 20% clay. Thus, the abstract notion of sediment components allows to deal with a wide range of physical properties using the same formalism. Sediments where natural remanence acquisition processes depend critically on clay or carbonate content are modeled in the same way as sediments where only changes in grain-size or composition of the magnetic minerals are relevant for remanence acquisition. The component grading can be arbitrarily arranged to account for nearly any factor, which is important for remanence acquisition behavior.

The function $c(z)$ completely describes the variation of sediment composition with depth. Such a variation may have several reasons. The source region of the sediment can change, transport mechanisms may vary in efficiency, sorting effects due to wind or current velocity can modulate grain-sizes. Moreover, salinity or water temperature and biological activity can change important sediment properties like carbonate content, clay mineralogy, or coagulation extent. Chemical variations may also modify weathering conditions and early diagenesis.

### 3.2.2.2 Uni-Causality and Sediment Linearity

Many of the above influences are not easily detected by standard geophysical or sedimentological methods. Yet, concentration variations of the components $P_i$ due to the above effects often are highly correlated: they all mirror the same environmental changes. At geologically and climatically "quiet" locations, it can be assumed that a predominant single environmental influence modulates *all* concentration changes. Therefore, the simplest *uni-causal* linear sediment model is obtained by assuming that $c(z)$ depends linearly on a single (environmental) master signal $s_1(z)$. For convenience, $s_1(z)$ is chosen to have zero mean and standard deviation of 1. Thus, for constant vectors $\bar{c}$ and $a$, the concentration vector is given by

$$c(z) = \bar{c} + s_1(z)a. \tag{3.1}$$

For component $P_i$, the average concentration then is $\bar{c}_i$ and its amplitude with respect to $s_1(z)$ is $a_i$ (see Fig. 3.5).

In more complex environments, it may be necessary to take into account several independent external influences. This can be obtained by using a *multi-causal* linear model with $k$ different master signals $s_1(z), \ldots, s_k(z)$. In practical applications these master signals and their respective influence can be reconstructed by principal component analysis of the investigated physical properties. The concentration vector in a multi-causal model is

$$c(z) = \bar{c} + \mathscr{A} \cdot s(z), \tag{3.2}$$

where $\mathscr{A}$ is a $k \times n$ matrix and $s(z) \in \mathbb{R}^k$. Nonlinear response to a single master signal $s(z)$ is mathematically equivalent to a multi-causal model. One way to see this is to interpret the $j$-th component of $s(z)$ in Eq. (3.2) as $s^j(z)$ to obtain a $k$-th order polynomial dependence of $c(z)$ upon $s(z)$. Nonlinear response is especially important when in the studied sedimentary sequence relevant phases vanish completely for some time or reach a state of saturation. However, the following investigation will focus on the case of uni-causal linear response.

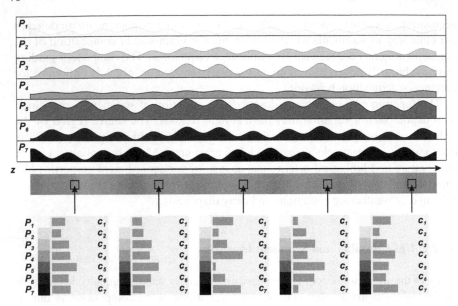

**Fig. 3.5** Linear sediment model with seven phases $P_1, \ldots, P_7$. At each depth $z$ the sediment is a mixture of these phases in varying concentrations $c_1(z), \ldots, c_7(z)$. If a single climate signal $s_1(z)$ controls all sediment phases, each concentration is an affine combination $c_i(z) = \bar{c}_i + a_i s_1(z)$

### 3.2.2.3 Linear Sediment Properties

A linear sediment property is a parameter $p(z)$ that depends linearly upon the concentration of the sediment components such that for some constant vector **p**,

$$p(z) = \langle \mathbf{p}, \mathbf{c}(z) \rangle = \sum_{i=1}^{n} p_i c_i(z), \tag{3.3}$$

where $\langle ., . \rangle$ is the standard inner product. A simple example is clay content. In this case, $p_i$ denotes the volume fraction of clay within the $i$-th component. The same can be done for the volume fraction of any sediment constituent and any extensive physical property. Accordingly, many magnetic parameters are linear sediment properties. If $p_i = \kappa_i$ denotes volume susceptibility of the $i$-th component, then $\kappa(z) = \langle \kappa, \mathbf{c}(z) \rangle$ is a linear sediment property if magnetostatic interaction between the components is negligible.

### 3.2.2.4 Linear Theory of Relative Paleointensity

The sediment model of the previous section is now applied to natural and synthetic remanence acquisition. This will help to understand the formation and reconstruction of relative paleointensity records and gives insight into their possible error sources.

To model NRM, it is assumed that each sediment component $P_i$ has its specific DRM acquisition constant $m_i$ representing magneto-mineral mobility, primary remanence intensity as well as relevant matrix properties. It is further assumed that the component acquires a DRM, which depends linearly on both, field $H(t(z))$ and mechanical activation $A(z)$. The activation term in the following will be assumed to be constant $A(z) = A_0$ over the whole sequence. Yet, it is kept in the formulae to emphasize the related uncertainty. In accordance with the above definitions the total NRM can be written as

$$
\begin{aligned}
\mathrm{NRM}(z) &= \sum_{i=1}^{n} \mathrm{NRM}_i(z) \\
&= A(z) H(t(z)) \sum_{i=1}^{n} m_i c_i(z) \\
&= A(z) H(t(z)) \langle \mathbf{m}, \mathbf{c}(z) \rangle.
\end{aligned} \tag{3.4}
$$

Thus, the susceptibility $m(z) = \langle \mathbf{m}, \mathbf{c}(z) \rangle$ of DRM acquisition with respect to external field and mechanical activation is a linear sediment property.

### 3.2.2.5 Synthetic Remanences as Linear Sediment Properties

In the same spirit it is assumed that each component $P_i$ contributes with its own proportionality factor $\gamma_i$ linearly to the normalizer $v(z)$. Accordingly, $v(z)$ is also a linear sediment property

$$
v(z) = \langle \gamma, \mathbf{c}(z) \rangle. \tag{3.5}
$$

By choosing a sufficiently fine grading of components, the linear model allows for arbitrary and independent grain-size dependence of both, DRM and normalizer. Therefore, considerable shift in lithology, mobility, or grain-size of the magnetic fraction can be present without corrupting the following conclusions.

### 3.2.2.6 Reconstruction of Relative Paleointensity

Under the above premises, $\mathrm{NRM}(z)$ and $v(z)$ can be calculated independently yielding

$$
\mathrm{NRM}(z) = A(z) H(t(z)) \left( \langle \mathbf{m}, \overline{\mathbf{c}} \rangle + s_1(z) \langle \mathbf{m}, \mathbf{a} \rangle \right) \tag{3.6}
$$

and similarly

$$
v(z) = \langle \gamma, \overline{\mathbf{c}} \rangle + s_1(z) \langle \gamma, \mathbf{a} \rangle. \tag{3.7}
$$

The latter equation can be solved for $s_1(z)$, if $\langle \gamma, \mathbf{a} \rangle \neq 0$. By substituting the result into (3.6) one obtains

$$\mathrm{NRM}(z) = A(z) H(t(z)) k_v (\overline{v} + v(z)), \tag{3.8}$$

where

$$k_v = \langle \mathbf{m}, \mathbf{a} \rangle / \langle \gamma, \mathbf{a} \rangle, \tag{3.9}$$

and

$$\overline{v} = k_v^{-1} \langle \mathbf{m}, \overline{\mathbf{c}} \rangle - \langle \gamma, \overline{\mathbf{c}} \rangle, \text{ when } k_v \neq 0. \tag{3.10}$$

It follows that not even in the linear first order multi-component model the 'normalized' NRM is necessarily proportional to the external field. If $\overline{v} \neq 0$ it still depends on $v(z)$:

$$\mathrm{NRM}(z)/v(z) = A(z) H(t(z)) k_v \left(1 + \frac{\overline{v}}{v(z)}\right). \tag{3.11}$$

The correct normalization has to take into account the constant offset $\overline{v}$ to the normalizer and is given by

$$\frac{\mathrm{NRM}(z)}{v(z) + \overline{v}} = k_v A(z) H(t(z)). \tag{3.12}$$

For constant $A(z)$ and known $\overline{v}$, Eq. (3.12) can be used to recover the paleofield.

Although more complicated than a simple division of NRM by the normalizer, Eq. (3.12) is remarkable because it doesn't require the normalizer to mirror the DRM acquisition with respect to grain-size as is requested by the current RPI paradigm for homogeneous sediments.

Also, the NRM is allowed to be carried by a complex mixture of phases, each responding individually to external field and mechanical activation.

In this respect, requiring the validity of the linear sediment model is a much weaker restriction upon the sediment record than the rigid homogeneity requirements of the current RPI paradigm.

### 3.2.2.7 Validity of the Classical Normalization Procedure

The above formulation allows to state the necessary conditions for the validity of the classical RPI normalization procedure. The $\mathrm{NRM}(z)$ actually is only proportional to $v(z)$ if $\overline{v} = 0$ in Eq. (3.10), which is equivalent to

$$\langle \mathbf{m}, \mathbf{a} \rangle \langle \gamma, \overline{\mathbf{c}} \rangle = \langle \gamma, \mathbf{a} \rangle \langle \mathbf{m}, \overline{\mathbf{c}} \rangle. \tag{3.13}$$

There are two independent possibilities for the validity of Eq. (3.13).

The first is that $\mathbf{m} = \sigma \gamma$ for some $\sigma \in \mathbb{R}$, which corresponds to the physical condition that each phase must contribute in exactly the same way to both, normalizer $\nu$ and NRM. This is related to the request of Levi and Banerjee (1976) that the normalizer must activate the same spectrum of magnetic particles which is responsible for NRM, an argument which has been further elaborated by Amerigian (1977) and King et al. (1983). A proposed test for this condition is to check for similarity of the demagnetization curves of NRM and normalizing remanence in function of sediment depth (Levi and Banerjee, 1976). In case of perfect parallelism, both demagnetization curves may vary with depth but in exactly the same way.

A second possibility is that $\bar{\mathbf{c}} = \lambda \mathbf{a}$ for some $\lambda \in \mathbb{R}$. The physical interpretation of this condition is that the relative composition of the sediment does not change with depth. All concentration changes of the sediment phases have to be perfectly parallel. Shifts in grain-size or composition must not occur. This is the request of perfect lithological homogeneity, which can be tested by comparing the downcore variation of the shape of ARM demagnetization curves (Levi and Banerjee, 1976) or more general by checking whether all concentration independent magnetic properties (shape of any isothermal magnetization loop, quantities like $H_{cr}$ or $H_c$) are invariant with depth.

Mathematically, both possibilities are independent. One of them alone suffices to guarantee the validity of the classical RPI normalization procedure. Therefore, a perfect normalizer should yield a good RPI record even for a less homogeneous sedimentary sequence, while even a mediocre normalizer should work well in a perfectly homogeneous lithology.

Of course Eq. (3.13) can be fulfilled by other combinations of $\mathbf{m}, \gamma, \bar{\mathbf{c}}, \mathbf{a}$, but these are unlikely to occur by chance.

Numerical modeling indicates that even if a linear sediment has a small nonzero bias $\bar{\nu}$ in Eq. (3.12), the classically inferred RPI record deviates only slightly from the correct field variation. In this case a better RPI estimate may be obtained by choosing an approximation of $\bar{\nu}$ which minimizes the average amplitude of the normalized record. This typically leads to only a minor amplitude reduction in comparison to the classically normalized record and may explain why the latter in inhomogeneous sediments often works by far more better than should be expected from considering the rigid prerequisites, which are usually demanded for its validity (Haag, 2000).

The above discussion relies on the linear uni-causal sediment model and therefore disregards highly nonlinear effects like VRM or diagenetic overprint. Perhaps, the main outcome of this study is that linearity and uni-causality already imply a sufficient regularity of the sedimentary sequence to make it useful for RPI reconstruction. Quality checks therefore should focus on testing to which degree the assumptions of uni-causality and linearity are fulfilled in a given sediment core and need not concentrate too much on idealistic homogeneity requests. Especially the choice of magnetic cleaning techniques should also be interpreted as a means to restrict RPI determination to linearly behaving sediment fractions.

# 3.3 Paleointensity for the Mesozoic and Cenozoic

## 3.3.1 Geomagnetic Field During the Mesozoic and Cenozoic

Paleomagnetic results allow for the investigation of long-term average features of the Earth's magnetic field. One long-standing basic assumption of paleomagnetism is that the geomagnetic field is on average an axial dipole, parallel to the Earth's axis of rotation. However, paleomagnetic data of the past million years show systematic departures from this simple model (Gubbins and Kelly, 1993). Persistent nondipolar terms of the geomagnetic field are required to fit paleomagnetic data of the last 5 Myr (McElhinny and Lock, 1996; Johnson and Constable, 1997). Pronounced features of the Johnson and Constable (1997) time-averaged field model are significant inclination anomalies of the geomagnetic field in the central Pacific and the Equatorial Atlantic. These features, however, might also be attributed to be an artifact due to poor distribution of and unrecognized errors within the data (Johnson and Constable, 1997; Carlut and Courtillot, 1998). Paleointensity data have also been used for such analysis (Kono and Hiroi, 1996, e.g.), but the paleointensity data set is, as yet, insufficient to obtain detailed models of time-averaged paleomagnetic field (Hatakeyama and Kono, 2002, e.g.). The available absolute paleointensity data, however, indicates a bimodal distribution of the Earth's average dipole moment during the geological past. Local investigations of paleomagnetic field strength indicate low values for mid Miocene rocks from the Canaries, approximately half of the present day field intensity (Leonhardt et al., 2000; Leonhardt and Soffel, 2002) and values in the range of the present day field around the last polarity reversal, the 0.78 Ma Matuyama-Brunhes transition (Valet et al., 1999). Such a geomagnetic field intensity difference between the Miocene and the Pleistocene is found globally (Juárez et al., 1998; Selkin and Tauxe, 2000; Shcherbakov et al., 2002; Heller et al., 2002). A bimodal distribution of intensities appears to be present in paleointensity data from the Mesozoic and Cenozoic (Shcherbakov et al., 2002; Heller et al., 2003). The low-field mode is characterized by values of about half the present day virtual dipole moment ($\approx 4 \times 10^{22}$ Am$^2$), the high field mode shows values as observed today ($8 \times 10^{22}$ Am$^2$).

### 3.3.1.1 Intensity Variations in Cretaceous Lava Flows: Case Studies from Franz-Josef-Land

Paleointensity investigations on Cretaceous volcanics provide important constraints on the duration and significance of the Mesozoic Dipole Low (MDL), a period dominantly characterized by a low-field state of the Earth's magnetic field. This feature, the MDL, was first identified by Bol'shakov and Solodovnikov (1981) and Prévot et al. (1990) and later confirmed by Tanaka et al. (1995) and Perrin and Shcherbakov (1997). Moreover, during this period the reversal rate displays a remarkable extreme: The Cretaceous is characterized by a 37 Ma long time interval of normal polarity and geomagnetic reversals are absent (Cretaceous Normal Superchron, CNS, for example, Opdyke and Channell, 1996). It has been proposed by Tarduno and Smirnov (2001) that this period is, in contrast to the rest of the Mesozoic, characterized by high values of the virtual geomagnetic dipole moment (VDM). They suggest that average paleointensities are highest when reversal rates are lowest, i.e. that the most stable geodynamo is most efficient and has the highest dipole moment. However, the abundance of reliable paleointensity data for this time interval is exceedingly sparse.

The archipelago of Franz-Josef-Land comprises 191 islands and constitutes the uplifted and dissected northern margin of the Barents Shelf in the arctic ocean. Specimens from unoriented samples of Cretaceous volcanics from Franz-Josef-Land, provided by A. Makareva and colleagues from the Polar Marine Geophysical Expedition (Lomonosov, Russia), were subjected to Thellier-type paleointensity investigations (MT4, Leonhardt et al., 2004a). Confirmed by in-depth rock-magnetic studies and ore microscopy, high quality paleointensity data was obtained in a preliminary study. The obtained paleointensity values were used to calculate the VDM in order to allow comparison to published data. As the paleolatitude of Franz-Josef-Land is basically undetermined a conservative approach was adopted: The VDM was calculated assuming inclinations between 70° and 90° (corresponding to paleolatitudes between 53° and 90°). Those values yield a high and low estimate for the true VDM. Although these values have to be considered as preliminary and a rough estimate, an interesting pattern is readily identified. The VDM results of radiometrically dated samples (Fig. 3.6, red boxes) agree mostly very well with published data.

Moreover, the data indicate that the Earth's magnetic field prior to and at the beginning of the CNS is characterized by an in average low virtual dipole moment

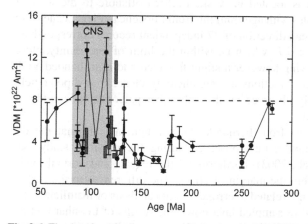

**Fig. 3.6** The data for this plot was compiled using the BOROKPINT.MDB database (Shcherbakov et al., 2002), http://wwwbrk.adm.yar.ru/palmag/index.html, selection criteria: only Thellier–Thellier method, mean of at least 5 cooling units, no data obtained from submarine basaltic glass). Data published recently not included in the database was added. *Red boxes* indicate VDM values estimated (details are given in the text) on the basis of the results obtained so far. *The gray shaded area* represents the Cretaceous Normal Superchron (CNS)

(VDM) with values around $2 - 5 \times 10^{22}$ Am$^2$. Also, toward the end of this period low VDM values are evidenced ($\sim 4 \times 10^{22}$ Am$^2$).

### 3.3.1.2 The Time-Averaged Geomagnetic Field During the Neogene: Results from Brazil, São Tomé and Tenerife

The present day geomagnetic field in the equatorial Atlantic is characterized by a distinct inclination anomaly. Time-averaged geomagnetic field models of the last 5 Myr, based on paleomagnetic data, indicate that this anomaly might be a persistent feature of the geomagnetic field (e.g., McElhinny and Lock, 1996; Johnson and Constable, 1997; Carlut and Courtillot, 1998). In particular, Johnson and Constable (1997) found strong departures from the geocentric axial dipole (GAD) hypothesis in the Equatorial Atlantic. This anomaly heavily relies on a few paleomagnetic data sets. Strong inclination and declination anomalies are present in old paleomagnetic data from Fernando de Noronha, Brazil (Richardson and Watkins, 1967; Schult et al., 1986), which is located in the western equatorial Atlantic. Paleomagnetic data from the eastern equatorial Atlantic is characterized by particularly strong inclination anomalies (Piper and Richardson, 1972).

Those earlier results were predominately obtained in order to establish a continental based magnetostratigraphy. For time-average field analysis high precision in directional analysis is needed. One might assume that due to limited demagnetization techniques and analysis methods, the previously obtained paleomagnetic data is insufficiently accurate. In order to test this hypothesis and to extend the available data set in terms of directions and paleointensities, new paleomagnetic samples were obtained from both islands, São Tomé and Fernando de Noronha.

The island of São Tomé is located in western Africa offshore to the west of Gabun. Paleomagnetic samples with ages ranging from Pliocene to Pleistocene were investigated. The average mean direction of 27 independent records corresponds to Dec: 184.2°, Inc: 10.3°, $\alpha_{95}$: 6.7° which is, within the limit of uncertainty, close to an axial dipole field, showing however a slight far-sided and right-handed VGP position. The obtained mean directions are significantly closer to a dipolar field configuration than previously observed by the only other paleomagnetic study of Piper and Richardson (1972).

New paleomagnetic results from Fernando de Noronha indicate that the directional anomaly is far smaller than observed during the previous study, but still present (Leonhardt et al., 2003). Furthermore, the reverse and normal mean directions are antipodal within the margin of errors. The differences to the earlier investigations, however, are not related to erroneous data analysis or insufficient demagnetization because in all re-sampled lava flows, the results of Leonhardt et al. (2003) are similar to those of Richardson and Watkins (1967) and Schult et al. (1986). In fact, the difference between the previous study and the new results is linked to the extension of the data set in terms of time coverage and quantity of sites, and to the application of directional groups.

Although it is not possible to draw a global conclusion from the observations from one or two sampling location and, in particular, when only few independent directions and intensities are available, these results emphasize the importance of extending the paleomagnetic database, especially for regions presently represented by poor sampling coverage, in order to facilitate conclusive field models describing whether persistent long-term features of the geomagnetic field exist or not. For Fernando de Noronha and São Tomé, new paleomagnetic results suggest that the present day dipole anomaly is not a persistent local feature of the Earths magnetic field.

Another important aspect of the time-averaged geomagnetic field of the last million years is the change from the low-field state observed during the mid Miocene and the present day high field state. The presence of two preferred states of field strength require some kind of geomagnetic switching process between those states. Paleointensity determinations were conducted on samples from Fernando de Noronha (~3 Ma, Leonhardt et al., 2003) and Tenerife (6 Ma, Leonhardt and Soffel, 2006). The VDMs of the Pliocene rocks from Fernando de Noronha are similar for the reverse and normal state of the field and the overall mean value of $(7.5 \pm 2.6) \times 10^{22} Am^2$ is only slightly lower than the present day magnetic moment. This observation is in agreement with the findings of Heller et al. (2002) whose paleointensity database analysis points to a Pliocene dipole moment close to $8 \times 10^{22} Am^2$. Within the late Miocene rocks of the Teno massif, Tenerife, the average geomagnetic field intensity is approximately half of the present field intensity (Leonhardt and Soffel, 2006). The obtained average value is within the low-field level of the Earth's magnetic field (VDM: $4.9 \times 10^{22} Am^2$). This data supports the presence of a bimodal distribution of the geomagnetic dipole moment and indicates that the switch between a low-field state and a high-field state happened between 3 and 6 Myrs ago.

## 3.3.2 A Central South Atlantic Paleointensity Stack for the Late Quaternary

A recent compilation of RPI records from sedimentary sequences recovered by Ocean Drilling Program (ODP) cruises all over the world (Lund et al., 2006) shows a large data gap in the central South Atlantic.

In this region, the sedimentation environment is very complex. Partly because, the corrosive Antarctic bottom water together with oligotrophic conditions results in very low sedimentation rates, but also due to climatic variations of terrigenous influx, and movements of the subtropical front. This, probably not exhaustive, list of complications has the effect that no reliable sedimentary magnetic records from this region were available up to now.

During Meteor cruise M46/4, 29 sediment cores were recovered from this critical region close to the Mid-Atlantic Ridge across the subtropical front (Wefer and cruise participants, 2001).

Out of this collection, we selected eight most appropriate sediment series and combined them into a stratigraphic network (Hofmann et al., 2005). Because of the highly variable oceanographic regimes the eight investigated sediment cores have widely varying sedimentary compositions, which for the reconstruction of geomagnetic variation usually is a most unwanted feature. The cores can be sorted into three lithologic groups, which correspond to their geographic positions with respect to the subtropical front (STF). We find this grouping also in the obtained RPI records that show clear similarities between the cores of the same lithologic group, although they otherwise can differ considerably. However, the cores were taken from a region small enough to assume a spatially uniform external field, and we will exploit this fact for a lithologic correction of the individual RPI records. Moreover, we apply principal component analysis of NRM and rock magnetic records to detect and single-out signal components that relate to independent but concurrent environmental signals. Thereby, we can for each individual core identify a remanence fraction, which should give the optimal RPI estimate. First, the individual RPI records are then directly stacked, by using the arithmetic mean of the RPI records over the time interval of the last 300 ka. This defines a first RPI stack (SAS-300) for the central South-Atlantic is defined. Although several features of this stack are still clearly controlled by the strongly varying lithologies, many of its characteristics can be related to other RPI stacks, such as SINT-800 (Guyodo and Valet, 1999), Sapis (Stoner et al., 2002) or the RPI record from ODP-Site 1089 (Stoner et al., 2003). In a second step, we then use a refined relative paleointensity model to correct the stack for lithologic influences.

### 3.3.2.1 Sampling Area and Core Lithology

Before analyzing the magnetic records, we outline geography, sedimentary environment, and chrono-stratigraphy of the here investigated stratigraphic network as developed in Hofmann et al. (2005). All eight network cores are located west of the Mid-Atlantic ridge in the subtropical and subantarctic South Atlantic across the STF. Core locations, water depths, ages, and sedimentary environments are summarized in Fig. 3.7. At the STF, northern and southern water masses collide and create a characteristic productivity gradient with low sedimentation rates in the North, and higher sedimentation rates south of the STF (Hofmann et al., 2005; Reid, 1989; Talley, 1996; Stramma and Peterson, 1990; Stramma and England, 1999). Shipboard lithologic description and measurements of $\kappa$ and density confirm that the selected eight cores have continuous and undisturbed lithologies, which can be ordered into three groups (Wefer and cruise participants, 2001; Hofmann et al., 2005). The two southernmost cores GeoB 6405-6 and GeoB 6408-4 are influenced by the Antarctic bottom water (AABW) and lie below the calcium carbonate compensation depth (CCD). They are classed as diatom bearing nanofossil ooze. Because the nearby core GeoB 6407-1 was taken from a shallower water depth (3384 m), its sediment composition rather resembles cores GeoB 6421-2 and GeoB 6422-1 from the vicinity of the STF. All three cores of this second group lie above the CCD and

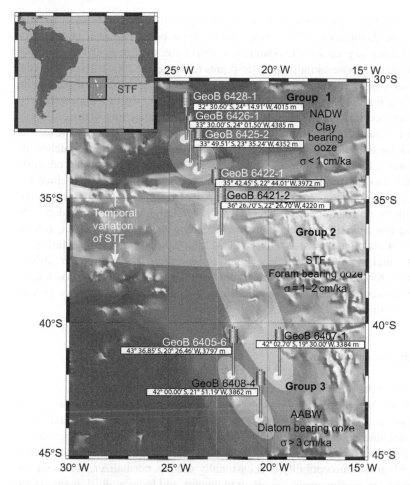

**Fig. 3.7** Location, water depth average sedimentation rates σ and grouping of the network cores. The inset shows the approximate current position of the STF as inferred from satellite productivity mapping. The band of temporal STF variation roughly represents our previous conclusion from δMn/Fe measurements

contain mainly foam bearing nanofossil ooze. The remaining cores, GeoB 6425-2, GeoB 6426-1, and GeoB 6428-1, come from the oligotrophic subtropical South Atlantic north of the STF. They reflect the sedimentation history of the nutrient poor NADW where, due to low carbonate or opal influx, clay minerals are dominant.

### 3.3.2.2 The Stratigraphic Network

A detailed age model for the three northern cores has been developed by Schmieder (2004) by correlating small scale characteristics of their magnetic susceptibility records to the Subtropical South Atlantic Susceptibility Stack (SUSAS) of von Dobeneck and Schmieder (1999). Since all northern cores recorded at least the Brunhes-Matuyama geomagnetic boundary at their base, the age model is additionally constraint by magnetostratigraphic tie-points (Schmieder, 2004). The age model for the remaining five cores was constructed using computer aided multi-parameter signal correlation as developed in Hofmann et al. (2005). In this approach precise inter-core correlation is achieved by synchronously matching high-resolution records of magnetic susceptibility $\kappa$, wet bulk density $\rho$, and X-ray fluorescence scans of elemental composition. Moreover, $\delta^{18}O$ records have been matched in the same process for the two cores where they are available. An optimal network correlation between all pairs of multi-parameter sequences has been obtained by controlling the correlation errors between core pairs $(A, B)$ and $(B, C)$ through comparison with the correlation $(A, C)$. The network correlation was then provided with an absolute age model by synchronously correlating the $\delta^{18}O$ and $\kappa$ records of GeoB 6408-4 and GeoB 6421-2 with SPECMAP, (Imbrie et al., 1984; Martinson et al., 1987) and SUSAS (von Dobeneck and Schmieder, 1999; Schmieder, 2004). Even though lithology varies considerably between the cores, it was thus possible to obtain high internal consistency, which ensures a relative median age error of less then 5 ka (Hofmann et al., 2005).

### 3.3.2.3 RPI Determination

As discussed in the previous section, to obtain an estimate of relative paleointensity (RPI), NRM needs to be normalized to correct for variations of remanence carrier concentration. The seemingly most reasonable normalizers are ARM and IRM, which are remanence based and often resemble the NRM in their demagnetization behavior. It is therefore assumed that they activate the same grains that carry the remanence. Magnetic susceptibility is occasionally used as normalizer, but its signal also encompasses paramagnetic, superparamagnetic, and large multi-domain grains that contribute little or nothing to NRM. Here we first follow the practical approach of using all three parameters as alternative normalizers, hoping that they either yield similar results, or that the resulting records hint toward a well-founded choice. In a second step we improve upon this common procedure by developing a correction method, which uses all network cores together to detect and suppress lithologic influences upon the relative paleointensity stack. Because PCA of all remanences (NRM, ARM, and IRM) resulted in a single dominating (>90 %) component in the 30 to 80 mT demagnetization interval, we also restricted the RPI determination to this interval by defining

$$NRM_{diff} := NRM_{30} - NRM_{80}, \qquad (3.14)$$

and similarly for $ARM_{diff}$ and $IRM_{diff}$. By comparing the RPI records obtained from the three different normalization parameters $ARM_{diff}$, $IRM_{diff}$, and $\kappa$, it turned out that, while $NRM_{diff}/ARM_{diff}$ and $NRM_{diff}/IRM_{diff}$ lead to very similar results, both

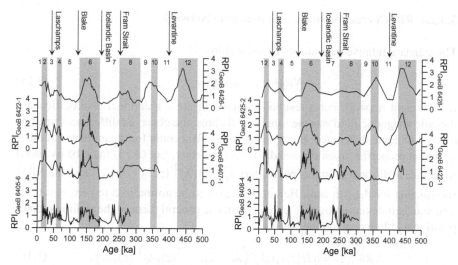

**Fig. 3.8** Individual relative paleointensity ($NRM_{diff}/ARM_{diff}$) records for the network cores. Lows in RPI correspond with common magnetic events. Slightly higher RPI values are found during glacial times (*marked gray*)

can considerably deviate from the susceptibility normalized RPI record $NRM_{diff}/\kappa$. Due to the closer resemblance between NRM and ARM demagnetization curves as compared to the IRM demagnetization curves, we will base the following analysis on the $NRM_{diff}/ARM_{diff}$ records that are collected in Fig. 3.8.

Most of the network cores display lows in paleointensity around 45 ka, corresponding to the Laschamp event, and around 100 ka, corresponding to the Blake event (Laj et al., 2000; Thouveny et al., 1990, 1993; Channell et al., 2000; Mazaud et al., 2002; Thouveny et al., 2004; Guillou et al., 2004). Two further lows in RPI at 190 ka and 250 ka are found in all cores except in the three northernmost, which have much lower sedimentation rate. The first low at 190 ka may correspond with theIcelandic Basin event (Channell, 1999), and the low at 250 ka with the Fram Strait event (Langereis et al., 1997; Nowaczyk and Baumann, 1992). In the northern cores, another RPI low at 400 ka, corresponding to the Levantine excursion (Thouveny et al., 2004; Langereis et al., 1997), is observed. In all cases, the inclination records are not anomalous. This probably is due to the low sedimentation rates, which smooth out short term field variations. However, it is also possible that the geomagnetic field, though weaker, did not noticeably change its local geometry. Throughout the network cores, slightly higher relative paleointensity values are found during glacial times (Fig. 3.8). In view of the above discussed shortcomings of the RPI methodology, this is more likely to be an artifact of insufficiently cancelled environmental signals, than an indication of climate-field coupling. To study this important topic of environmental influence more completely, the next paragraphs investigate the influences of compositional variations upon the NRM signal in the network cores.

### 3.3.2.4 RPI Correction in the Stratigraphic Network

The standard relative paleointensity estimation

$$NRM_i(z) = q_i H(z) c_i(z). \tag{3.15}$$

for the eight network cores resulted in quite different individual results. Because the paleomagnetic field was the same for all cores, the observed differences must be due to lithologic differences between the cores. To account for the influence of other sediment properties upon the standard paleointensity, Eq. (3.15) must be extended to include additional sediment parameters $\lambda_i^{(1)}(z), \lambda_i^{(2)}(z), \lambda_i^{(3)}(z), \ldots$ into the NRM formulation for the $i$-th core. Herein, $\lambda_i^{(j)}(z)$ is to be normalized to zero mean and unit standard deviation over all cores. The most general form of the sediment compositional influence upon the NRM is

$$NRM_i(z) = q H(z) c_i(z) f\left( \lambda_i^{(1)}(z), \lambda_i^{(2)}(z), \lambda_i^{(3)}(z), \ldots \right), \tag{3.16}$$

where $f$ is an arbitrary differentiable function, and $q$ now is a common absolute calibration constant valid for all cores. By assuming further that the compositional parameters $\lambda_i(z)$ have only a relatively weak influence in comparison to the external field, (3.16) can be linearized to yield

$$NRM_i(z) = q H(z) c_i(z) \left( 1 + \alpha_1 \lambda_i^{(1)}(z) + \alpha_2 \lambda_i^{(2)}(z) + \alpha_3 \lambda_i^{(3)}(z), \ldots \right). \tag{3.17}$$

This expression can be simplified further, if a single parameter $\lambda_i(z)$ is predominant in this expansion. Then, the other parameters can be neglected, and the common external field variation $H_i(z)$ for each core is given by

$$NRM_i(z) = q H_i(z) c_i(z) \left( 1 + \alpha \lambda_i(z) \right). \tag{3.18}$$

Because the $\lambda_i$ have a common normalization, $\alpha$ is a quantitative measure of the influence of $\lambda$ upon field recording in the network area. Using the stratigraphic network, it is possible to determine $\alpha$ by minimizing the total squared distance between the single-core field estimates $H_i(z)$. This method is now used to calculate the efficiencies $\alpha$ for different sediment properties in order to 1) determine the most influential one, and 2) to correct the single-core records using this property and $\alpha$.

### 3.3.2.5 Determining $\alpha$ by Optimizing the Correlation Between the $H_i(z)$

The age interval covered by all network cores is 270 ka, and the following investigations are confined on this age interval. Due to its very low sedimentation rate (0.4 cm/ka) and its too low resolution in this time interval the northernmost core GeoB 6428-1 has been excluded.

Figure 3.9 explains the network determination of $\alpha$ for the grain size sensitive parameter $\lambda = $ ARM/IRM. The initial state (a), where $\alpha = 0$ corresponds to the

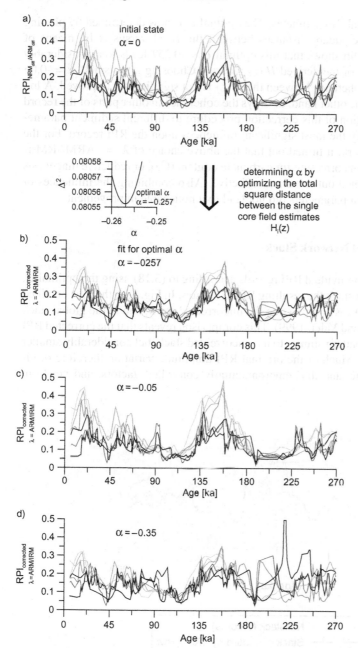

**Fig. 3.9** Correction procedure to determine the optimal value of $\alpha$ in Eq. (3.18). The individual RPI records (**a**) are linked by the correction procedure through their common lithological parameter $\lambda$. The strength of this link is related to the modulus of *alpha*. Choosing $\alpha$ too small (**c**) leads to negligible changes of the individual records. A too large $\alpha$ over estimates the lithologic influence and destroys the paleofield signal (**d**). The optimal choice of $\alpha$ (**b**) introduces a lithologic correction, which leads to a most coherent paleointensity estimate

uncorrected individual RPI estimates. The optimal $\alpha$ is now determined by plotting the sum $\Delta^2$ over the square distances between the $H_i(z)$ versus $\alpha$ by means of Eq. (3.18). Figure 3.9b shows that this optimal $\alpha = -0.257$ leads to a considerably improved coherence of the derived $H_i(z)$ records. Choosing a too small $\alpha = -0.05$ as in Fig. 3.9c, the coherence between the $H_i(z)$ is not significantly improved. A too large $\alpha = -0.35$, on the other hand, destroys the coherence in other parts of the record (Fig. 3.9d). Application of this correction procedure to different sediment parameters allows to identify the most significant influences upon the RPI record. For the South Atlantic network, it turned out that the above choice of $\lambda$ = ARM/IRM is most influential. Even correcting for carbonate content (Ca), or reductive diagenesis indicators (Fe/k) turned out to be less effective. Also combining the influences of two or more sediment parameters did not lead to a noticeable improvement.

### 3.3.2.6 A Corrected Network Stack

After correcting the individual RPI records according to (3.18) using the optimal $\alpha$, it is possible to again stack the resulting RPI estimates. In Fig. 3.10 the thus obtained stack of corrected RPI is compared to the uncorrected stack, and to the global stack SINT-800 (Guyodo and Valet, 1999). This comparison reveals that the corrected RPI has a characteristic pattern similar to the uncorrected stack, but considerably smaller amplitude variation. Much of the original RPI amplitude variation therefore originated from lithologic, and thus environmentally controlled, factors, and not from

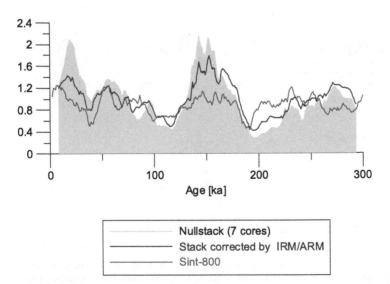

**Fig. 3.10** Comparison between the uncorrected RPI stack (Nullstack), the optimally corrected RPI stack, and SINT-800 (Guyodo and Valet, 1999). Nearly throughout the investigated time-interval, the correction moves the intensity of the uncorrected stack toward the SINT-800 intensity. The correction substantially reduces the predicted paleointensity amplitudes

the paleofield. While environmental influence upon RPI records via ARM/IRM was long since suspected, this study for the first time gives a quantitative verification of this effect. Near 20 ka the relative field amplitude is corrected from 2.1 to 1.4, which means that it was originally overestimated by a factor of 1.5. However, there are also time intervals where the corrected RPI estimate is higher than the uncorrected. This is mainly the case before 160 ka. Interestingly, the correction nearly everywhere moves the RPI stack toward the—completely independent—SINT-800 record, even though details of the pattern remain significantly different.

### 3.3.3 Environmental Influence upon Relative Paleointensity Records

The correction removed a relatively large environmental influence, which was contained in the uncorrected stack. This leads to the question how this environmental signal looks like, and whether it could have been detected by other means. The removed environmental signal is best seen in the ratio of corrected to uncorrected stack in Fig. 3.11. Comparison with SPECMAP and the insolation signal of July 21 at 65°N indicates that it is indeed a global climate signal that was present, though not directly visible, in any of the individual RPI records.

The fact that the detected climate signal is a global signal has important ramifications for the interpretation of sedimentary relative paleointensity records. First, it poses a problem for global stacks that intend to remove environmental influence by averaging a wide variety of globally distributed RPI records. This approach,

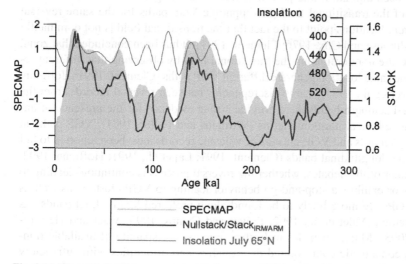

**Fig. 3.11** The ratio between uncorrected and corrected RPI stacks (*blue*) reveals a clear influence of global climate signals like ice-cover (SPECMAP) or critical insolation (July 21 at 65°N)

however, cannot succeed when the environmental signal itself is global. Second, in the light of our finding, the presence of a hidden global environmental signal appears highly likely in every individual RPI record. It is therefore doubtful whether valid conclusions about possible physical connections between geomagnetic field generation and global climate can be drawn on the basis of sedimentary RPI records.

## 3.4 Geomagnetic Reversals and Excursions

The switch between the two preferred states of the Earth's magnetic field, the reverse and the normal state, is commonly referred to as a polarity reversal. These reversals occurred often but irregular during the Earth's history, approximately once every 250 kyr during the last million years. The last full polarity reversal, the Matuyama-Brunhes transition occurred ~780 kyrs ago. While before and after such polarity reversals, the Earth magnetic field is essentially an axial dipole, the details of its transitional structure are still largely unknown. The temporal evolution of the geomagnetic field geometry during a reversal is intimately linked to the mechanism of field generation in the Earth's core. To better understand that mechanism, paleomagnetic records that provide inclination $I$, declination $D$, and sometimes intensity $F$ of the local transitional geomagnetic field are analyzed. A paleomagnetic determination of $I$ and $D$, is often transformed in its uniquely determined virtual geomagnetic pole (VGP), which denotes the position of the magnetic South pole of a geocentric dipole which would have produced the locally observed $I$ and $D$. A stratigraphic sequence of transitional paleomagnetic data results in a local VGP path which would correspond to the rotation of this geocentric dipole if the field had in fact been dipolar. Using paleomagnetic data it is possible to infer structural properties of the transitional field. Incompatible VGP paths for the same reversal from different locations point to the fact that the transitional field is not dominantly dipolar (Hillhouse and Cox, 1976; Clement, 1991). It has been concluded that zonal components are not dominating the transitional field by observing that different reversals at the same location generate different VGP paths (Clement, 1991; Hoffman, 1981; Theyer et al., 1985). Large discrepancies between reversal records from the northern and southern hemisphere provide strong evidence for the presence of antisymmetric spherical harmonic terms (Williams and Fuller, 1981). VGP-paths in sedimentary records and VGP-clusters of volcanic records have been reported to fall into preferred longitudinal bands (Clement, 1991; Laj et al., 1991; Hoffman, 1992, 1996). Neither of the debates, whether the reversal process is continuous leading to VGP paths or exhibits a stop-and-go behavior leading to VGP clusters, as well as whether VGPs are more likely to be found along preferred longitudinal bands has yet been settled (Valet et al., 1992; Prévot and Camps, 1993; Valet and Herrero-Bervera, 2003). Major obstacles to further progress are sparsity of available transitional paleomagnetic data, limited data quality, and difficulties with sufficiently accurate age determination (e.g., Roberts and Winklhofer, 2005). Suitable paleomagnetic records of polarity reversals obtained in volcanic rocks are discussed by

Heunemann et al. (2004), Nowaczyk and Knies (2000), Nowaczyk et al. (2003), and Leonhardt et al. (2000, 2002). In the following some of these records will be discussed, and it will be shown what information regarding geomagnetic field structure, field evolution, and reversal duration can be extracted. Furthermore, it will be scrutinized whether a reliable reconstruction of the spherical harmonic description of the global geomagnetic field throughout a reversal can be inferred from local paleomagnetic records.

### 3.4.1 Tertiary Field Reversals

In the mid-Miocene shield basalts of Gran Canaria, a paleomagnetic investigation has been carried out on a series of 87 volcanic lava flows (Leonhardt et al., 2002; Leonhardt and Soffel, 2002). The sequence covers a reversal of the Earth's magnetic field from reverse to normal. Intermediate directions of the field are recorded in 34 lava flows. From radiometric age constraints this transition corresponds very likely to a ~14.1 Ma field reversal. During the reversal the virtual geomagnetic poles show an accumulation in three consecutive areas: southeast of South America, east of India, and in the central Pacific (Fig. 3.12). In addition to the reversal, three excursions are recorded in the sequence, two preceding and one succeeding the reversal. The intermediate directions of the excursion just before and after the transition have the same VGP positions as the first and the last accumulation zone, respectively.

The remarkable returns to the same transitional positions near South America and the central Pacific indicate preferred locations of the VGP during the transitional stages. Only the South America cluster can be correlated to VGP clusters observed

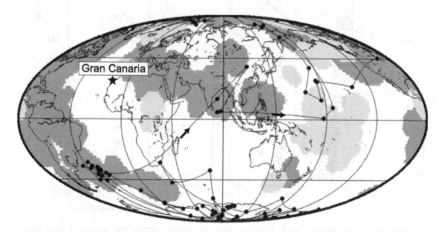

**Fig. 3.12** VGP movement of the El Paso sequence. The *light shaded* regions show ultralow-velocity zones (ULVZ) in the lowermost mantle, the *dark shaded* regions mark areas where these seismic features are not detected. A relation between VGP location and regions without ULVZ is visible only for the cluster near South America and the pole positions near India

in other reversals of the last 10 Myr and to lower mantle seismic anomalies. The VGPs in the Pacific track instead along the antipode of the site longitude. This indicates that mechanisms other than lower mantle heterogeneities additionally affect the transitional record of the sampled sequence and zonal harmonics are plausible candidates. In general, the obtained field intensities are lower than expected for the mid Miocene (Fig. 3.13).

This observation is very likely related to a long-term reduction of the field close to transitions. Very low paleointensities with values <5 μT were obtained during an excursion, preceding the actual transition, and also close to significant changes of the local field directions. These are interpreted as nondipolar components becoming dominant for short periods and provoking a rapid change of field directions. During

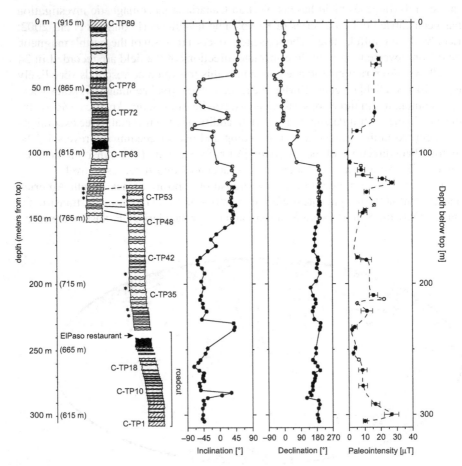

**Fig. 3.13** Inclination, declination, and absolute paleointensity across the sampled sequence (Leonhardt and Soffel, 2002). Within the inclination and declination diagrams, *solid circles* represent the data of profile A and *open circles* the data of profile B. *Open circles* within the intensity diagram indicate lavas with only one successful paleointensity determination. The South American VGP cluster ranges from flow C-TP47 to C-TP61

the transition, 15 successive lava flows recorded similar local field directions corresponding to a cluster of virtual geomagnetic poles close to South America. Chronologically, within this cluster the paleointensity increases from about 9 μT to 28 μT followed by a decrease back to approximately 9 μT This variation of paleointensity between lava flows with similar direction indicates independent records of the geomagnetic field and gives strong evidence for a stable transitional state, which lasted a significant period of time. The duration of 4500 years was estimated for this reversal using the secular variation rate recorded by the lavas.

Transitional geomagnetic field behavior was also observed on the Hawaiian islands of Maui and Lanai. Twenty-nine noncontiguous lava flows were sampled on Lanai. The lava flows formed during the late Matuyama polarity chron about 1.6 Myr ago and recorded the Gilsá geomagnetic polarity excursion. The obtained Ar/Ar ages of 1.6 Ma for the Lanaian lavas are significantly older than ages previously obtained for this island. These new ages are, however, in agreement with the age progression of the Hawaiian hotspot. The paleomagnetic record over the sampled succession is characterized by reversed and intermediate directions. Paleointensities are generally very low, about half of the present day field intensity, dropping to values of ~6 μT during transitional field states. Transitional VGPs from Lanai are situated first near the west coast of South America and then switch to offshore western Australia (Fig. 3.14).

Normal polarity is never reached in the sampled succession. Therefore, the Gilsá is referred to as an excursion of the geomagnetic field. Transitional VGP's located close to Australia accompanied by relatively low paleointensity are found in most short geomagnetic events/excursions during the Matuyama polarity chron, for example, the Cobb Mountain subchron, the Punaru event, and the Matuyama-Brunhes precursor, suggesting very similar physical processes likely related to lower mantle heterogeneities during these geomagnetic events. Approximately 800 kyr after the Gilsá excursion, lava flows on Maui recorded the last reversal, the Matuyama/Brunhes transition. Fourteen lava flows were sampled, which recorded the pre- and post-transitional behavior of the Earth's magnetic field. The geomagnetic

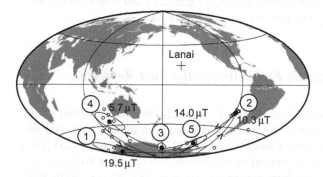

**Fig. 3.14** Virtual geomagnetic poles (VGPs) for individual lavas (*circles*) and the five directional units (*solid circles* with 95%-error ellipse) for Lanai. Also shown are the average paleointensity values for the directional units

field intensity is low prior to the reversal, approximately 8 μT, and increases strongly afterwards up to ~63 μT. These values of about twice the present field intensity of Hawaii after the transition emphasize a strong asymmetry between pre- and post-transitional fields, which is particularly strong in the region of Hawaii.

### 3.4.2 The End of the Kiaman Superchron

The behavior of the Earth's magnetic field during the Mesozoic and Late Paleozoic, or more precisely between 86 and 276.5 Ma (Shcherbakov et al., 2002), is of particular interest. During this time interval, also referred to as the Mesozoic dipole low (MDL), the VDM seems to have been significantly reduced compared to today's values. This apparent difference rises the question whether field reversal characteristics as observed in paleomagnetic data are different as well, for example, due to an even more pronounced influence of multipole components.

An approximately 250 Ma old reversed to normal transition of the Earth's magnetic field was sampled in 86 volcanic lava flows of the Siberian trap basalts, North Siberia, Russia (Gurevitsch et al., 2004; Heunemann et al., 2004). Transitional directions were obtained from 20 flows (Fig. 3.15).

During the reversal a clustering of the VGPs is observed (15 flows). Paleointensity estimates suggest that this feature is not an artefact due to rapid flow emplacement since the directional cluster is associated with a well-defined increase in paleointensity from 6 μT to 13 μT. Subsequently, the next VGPs move toward the pole position of normal polarity. Departing in a sudden movement from normal polarity the VGPs form a second directional cluster comprising the results of 14 flows. This feature is interpreted as a post-transitional excursion but lacks the characteristic intensity variation recorded during the first transitional cluster. The rest of the section (41 flows) is of normal polarity. The characteristic features of this reversal, low intensities and directional clustering during the reversal and an excursion shortly after the reversal, were also observed in records of polarity transitions of younger age (e.g., Hoffman, 1992; Leonhardt et al., 2002; Leonhardt and Soffel, 2002). This suggests that the underlying reversal processes were similar.

### 3.4.3 The Global Magnetic Field During the Last Reversal

Many researchers have emphasized the need for a reliable reconstruction of the spherical harmonic description of the geomagnetic field throughout a reversal and even called this the ultimate aim of transition studies. Leonhardt and Fabian (2007) presented a Bayesian inversion method to reconstruct the spherical harmonic expansion of this transitional field from paleomagnetic data. This is achieved by minimizing the total variational power at the core-mantle boundary (CMB) during the transition under paleomagnetic constraints. This measure makes minimal physical

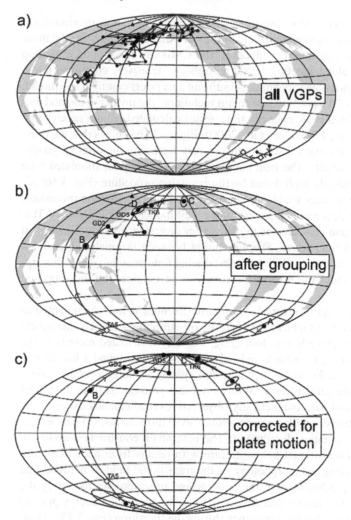

**Fig. 3.15** Movement of the VGPs across the studied sections Talnakh, Listvjanka, and Icon/Abagalakh (see Heunemann et al. (2004) for a detailed description). The site location is denoted by a triangle. (**a**) all VGPs, the results of the Talnakh profile, parallel to the bottom of the Listvjanka section, are denoted by open diamonds. (**b**) Shown as black dots are directional groups of statistically similar directions with their associated $d_m$ and $d_p$ confidence limits and the VGPs of individual flows not belonging to one of the groups. The lowest part is characterized by reversed polarities (group A and two flows of the Talnakh section) and then reaches a stable transitional cluster (Group B, 15 flows). The following four flows (GD2 to GD5) move toward normal polarity. The first flows (TK7 and TK6) of the Icon/Abagalakh section have directions typical for Early Triassic normal polarity, but then an excursion to more easterly pole positions is observed (group C, 14 flows) before reaching again the normal pole (group D). (**c**) The same VGPs as in (**b**) after correction for Permo-Triassic paleogeography. Continents in (**a**) and (**b**) are plotted in their present positions

assumptions about the geodynamo process itself and leads to a unique maximum entropy fit assuming that a change in the field at the core-mantle boundary is the more likely, the less energy variation it produces at the CMB. By the proposed method several high quality paleomagnetic reversal records are iteratively combined into a single geometric reversal scenario without assuming an a priori common age model. To test the suitability of the proposed inversion technique, the inversion method was applied to four simulated paleomagnetic time series calculated from a geodynamo model (Glatzmaier et al., 1999; Coe et al., 2000). Figure 3.16a shows that the energetic evolution of the inversion result compares very well with the original geodynamo model, even in details. The field components of the original simulated time series are almost identically reproduced by the inversion procedure (Fig. 3.16b,c). The field components of sites which are not used for the reconstruction procedure, referred to as independent records, are also very well modeled (Fig. 3.16d,e). This underlines the predictive character of the Bayesian inversion procedure even if it is based only on four time series. The agreement between original and modeled energetic evolutions and the capability of reproducing field vectors proves that significant information about the reversal process can be obtained by the proposed technique.

Being the last full polarity reversal, Matuyama-Brunhes is best documented in terms of number and geographical distribution of sampling sites. By iteratively applying the inversion procedure to four geographically distributed records of the last geomagnetic reversal, Leonhardt and Fabian (2007) constructed a low degree global field evolution model which then is tested against all other available paleomagnetic records of the same reversal. It turns out that predictions from the model are consistent with most paleomagnetically observed reversal characteristics of the Matuyama-Brunhes transition like (a) transitional VGP movements, (b) paleointensity variation, and (c) duration distribution. The obtained reconstruction invites to compare the inferred transitional field structure with results from numerical geodynamo models regarding the morphology of the transitional field. It is found that radial magnetic flux patches form at the equator and move polewards during the transition (Fig. 3.17). Our model indicates an increase of nondipolar energy prior to the last reversal and a nondipolar dominance during the transition (Fig. 3.17). Thus, the character and information of surface geomagnetic field records is strongly site dependent. The reconstruction also offers new answers to the question of existence of preferred longitudinal bands during the transition and to the problem of reversal duration. Different types of directional variations of the surface geomagnetic field, continuous or abrupt, are found during the transition. Two preferred longitudinal bands along the Americas and East Asia are not predicted for uniformly distributed sampling locations on the globe. Similar to geodynamo models with CMB heat flux derived from present day lower mantle heterogeneities (Glatzmaier et al., 1999), a preference of transitional VGPs for the Pacific hemisphere is found. The paleomagnetic duration of reversals shows not only a latitudinal, but also a longitudinal variation. Even the paleomagnetically determined age of the reversal varies significantly between different sites on the globe.

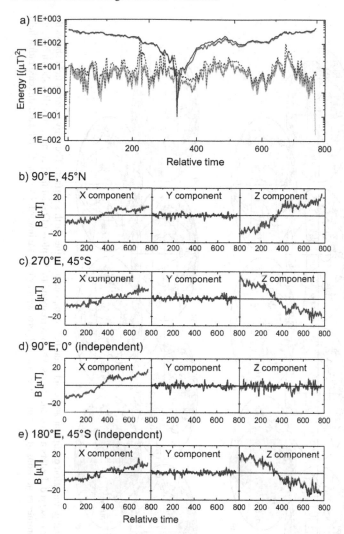

**Fig. 3.16** Application of the Bayesian inversion method (Leonhardt and Fabian, 2007) to four synthetic paleomagnetic records for sampling locations at (45°N, 90°E), (45°N, 90°W), (45°S, 90°E), and (45°S, 90°W) from the numerical geodynamo solution g2 (Glatzmaier et al., 1999; Coe et al., 2000). (**a**) Comparing the energy variation of dipolar (*solid lines*) and nondipolar (*dashed lines*) components of the original dynamo solution (*blue*) with our reconstructed inverse model (*red*) shows good agreement in both, long-term and short-term features. (**b,c**) The **X**, **Y**, and **Z** components of the synthetic records used for the inversion procedure are almost identically reproduced by the inverse model. (**d,e,**) Field components of independent records, i.e. records not used for the inversion procedure, are also very well matched by the inversion results. These results demonstrate the reliability of the Bayesian inversion

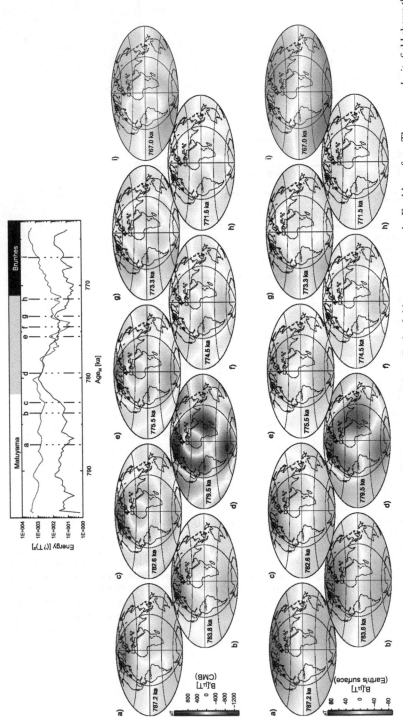

**Fig. 3.17** *Top:* Reconstruction of temporal energy variation of dipolar and nondipolar field components at the Earth's surface. The *gray polarity* field shows the time interval where transitional directions are observed. Letters a-i indicate times which correspond to the plots of radial field components ($B_r$) shown beneath. CMB and Earth's surface are displayed in Hammer projections of the same size in a-i. *Orange shades* indicate inward, *blue shades* outward-directed radial field.

# References

Aitken, M. J., Allsop, A. L., Bussell, G. D., and Winter, M. B. (1988). Determination of the intensity of the Earth's magnetic field during archaeological times: reliability of the Thellier technique. *Rev. Geophys.*, 26:3–12.

Amerigian, C. (1977). Measurement of the effect of particle size variation on the detrital remanent magnetization to anhysteretic remanent magnetization ratio in some abyssal sediments. *Earth Planet. Sci. Lett.*, 36:434–442.

Bol'shakov, A. and Solodovnikov, G. (1981). Intensity of the geomagnetic field in the last 400 million years. *Doklady Earth Sci. Sections*, 260:24–27.

Carlut, J. and Courtillot, V. (1998). How complex is the time-averaged geomagnetic field over the past 5 Myr? *Geophys. J. Int.*, 134:527–544.

Channell, J. (1999). Geomagnetic paleointensity and directional secular variation at Ocean Drilling Program (ODP) Site 984 (Bjorn Drift) since 500 ka: Comparisons with ODP Site 983 (Gardar Drift). *J. Geophys. Res.*, 104(B10):22937–22951.

Channell, J., Valet, J.-P., Hodell, D., and Charles, C. (2000). Geomagnetic paleointensity from late Brunhes-age piston cores from the subantarctic South Atlantic. *Earth Planet. Sci. Lett.*, 175:145–160.

Clement, B. (1991). Geographical distribution of transitional VGPs: Evidence for nonzonal equatorial symmetry during the Brunhes-Matuyama geomagnetic reversal. *Earth Planet. Sci. Lett.*, 29:48–58.

Coe, R. (1967a). Paleo-intensities of the Earth's magnetic field determined from Tertiary and Quaternary rocks. *J. Geophys. Res.*, 72:3247–3262.

Coe, R. S. (1967b). Palaeointensity of the Earth's magnetic field determined from Tertiary and Quaternary rocks. *J. Geophys. Res.*, 72:3247–3262.

Coe, R. S. (1974). The effect of magnetic interactions on paleointensity determinations by the Thelliers' method. *J. Geomag. Geoelectr.*, 26:311–317.

Coe, R. S., Hongre, L., and Glatzmaier, G. A. (2000). An examination of simulated geomagnetic reversals from a palaeomagnetic perspective. *Philos. Trans. R. Soc. Lond.*, 358:1141–1170.

Coe, R. S., Singer, B. S., Pringle, M. S., and Zhao, X. (2004a). Matuyama-Brunhes reversal and Kamikatsura event on Maui: paleomagnetic directions, $^{40}$Ar/$^{39}$Ar ages and implications. *Earth Planet. Sci. Lett.*, 222:667–684.

Coe, R. S., Rijsager, J., Plenier, G., Leonhardt, R., and Krása, D. (2004b). Multidomain behaviour durinh Thellier paleointensity experiments: Results from the 1915 Mt. Lassen flow. *Earth. Planet. Sci. Lett.*, 247:141–153.

Dekkers, M. J. and Boehnel, H. (2006) Reliable absolute paleointensities independent of magnetic domian state. *Earth Planet. Sci. Lett.*, 248:507–5016.

Dodson, M. H. and McClelland-Brown, E. (1980). Magnetic blocking temperatures of single-domain grains during slow cooling. *J. Geophys. Res.*, 85:2625–2637.

Dunlop, D. (1973). Superparamagnetic and single-domain threshold sizes in magnetite. *J. Geophys. Res.*, 78:1780–1793.

Dunlop, D. J. and Özdemir, Ö. (2000). Effect of grain size and domain state on thermal demagnetization tails. *Geophys. Res. Lett.*, 27:1311–1314.

Fabian, K. (2001). A theoretical treatment of paleointensity determination experiments on rocks containing pseudo-single or multi domain magnetic particles. *Earth Planet. Sci. Lett.*, 188:45–58.

Fabian, K. and Shcherbakov, V. P. (2004). Domain state stabilization by iterated thermal magnetization processes. *Geophys. J. Int.*, 159:486–494.

Fox, J. M. W. and Aitken, M. J. (1980). Cooling-rate dependency of thermoremanent magnetisation. *Nature*, 283:462–463.

Glatzmaier, G. A., Coe, R. S., Hongre, L., and Roberts, P. H. (1999). The role of the Earth's mantle in controlling the frequency of geomagnetic reversals. *Nature*, 401(6756):885–890.

Gubbins, D. and Kelly, P. (1993). Persistent patterns in the geomagnetic field over the past 2.5 myr. *Nature*, 365:829–832.

Guillou, H., Singer, B., Laj, C., Kissel, C., Scaillet, S., and Jicha, B. (2004). On the age of the Laschamp geomagnetic excursion. *Earth Planet. Sci. Lett.*, 227:331–343.

Gurevitsch, E. L., Heuneman, C., Radk'ko, V., Westphal, M., Bachtadse, V., Pozzi, J. P., Feinberg, H. (2004). Paleomagneism and magnetostratigraphy of the Permian-Triassic Sibirian trap basalts. *Tectonophysics*, 379:211–226.

Guyodo, Y. and Valet, J.-P. (1999). Global changes in the intensity of the Earth's magnetic field during the past 800 kyr. *Nature*, 399:249–252.

Haag, M. (2000). Reliability of relative palaeointensities of a sediment core with climatically-triggered strong magnetisation changes. *Earth Planet. Sci. Lett.*, 180:49–59.

Halgedahl, S. and Fuller, M. (1980). Magnetic domain observations of nucleation processes in fine particles of intermediate titanomagnetite. *Nature*, 288:70–72.

Harrison, C. G. A. (1966). The paleomagnetism of deep-sea sediments. *J. Geophys. Res.*, 71:3033–3043.

Harrison, C. G. A. and B. L. K. Somayajulu (1966). Behaviour of the Earth's magnetic field during a reversal. *Nature*, 212:1193–1195.

Hatakeyama, T. and Kono, M. (2002). Geomagnetic field model for the last 5 my: time-averaged field and secular variation. *Phys. Earth Planet. Inter.*, 133:181–215.

Heller, R., Merrill, R. T., and McFadden, P. L. (2002). The variation of intensity of the Earth's magnetic field with time. *Phys. Earth Planet. Inter.*, 131:237–249.

Heller, R., Merrill, R. T., and McFadden, P. L. (2003). The two states of paleomagnetic field intensities for the past 320 million years. *Phys. Earth Planet. Inter.*, 135:211–223.

Henkel, O. (1964). Remanenzverhalten und Wechselwirkung in hartmagnetischen Teilchenkollektiven. *Phys. Stat. Sol.*, 7:919–929.

Heunemann, C., Krása, D., Soffel, H. C., Gurevitch, E., and Bachtadse, V. (2004). Directions and intensities of the Earth's magnetic field during a reversal: results from the Permo-Triassic Sibirian trap basalts, Russia. *Earth Planet. Sci. Lett.*, 218:197–213.

Hillhouse, J. and Cox, A. (1976). Brunhes-Matuyama polarity transition. *Earth Planet. Sci. Lett.*, 29:51–64.

Hoffman, K. A. (1981). Palaeomagnetic excursions, aborted reversals and transitional fields. *Nature*, 294:67–69.

Hoffman, K. A. (1992). Dipolar reversal states of the geomagnetic field and core mantle dynamics. *Nature*, 359:789–794.

Hoffman, K. A. (1996). Transitional paleomagnetic field behavior: Preferred paths or patches? *Surv. Geophys.*, 17:207–211.

Hofmann, D., Fabian, K., Schmieder, F., and Donner, B. (2005). A stratigraphic network from the subtropical and subantarctic South Atlantic: Multiparameter correlation of magnetic susceptibility,bulk density, X-ray fluorescence measurements and $\delta^{18}O$. *Earth Planet. Sci. Lett.*, 240:694–709.

Imbrie, J., Hays, J. D., Martinson, D. G., McIntyre, A., Mix, A. C., Morley, J. J., Pisias, N. G., Prell, W. L., and Shackleton, N. J. (1984). The orbital theory of Pleistocene climate: Support from a revised chronology of the marine $\delta^{18}O$ record. In Berger, A., Imbrie, J., Hays, J., Kukla, G., and Saltzman, B., editors, *Milankovitch and Climate: Understanding the Response to Orbital Forcing*, pp. 269–305. Reidel Publishing, Dordrecht.

Johnson, C. L. and Constable, C. G. (1997). The time-averaged geomagnetic field: global and regional biases for 0-5 Ma. *Geophys. J. Int.*, 131:643–666.

Johnson, E. A., Murphy, T., and Torreson, O. W. (1948). Pre-history of the Earth's magnetic field. *Terr. Mag. Atmos. Elec.*, 53:349–372.

Johnson, H. P., Kinoshita, H., and Merrill, R. T. (1975). Rock magnetism and paleomagnetism of some North Pacific deep-sea sediments. *Geol. Soc. Am. Bull.*, 86:412–420.

Juárez, M. T., Tauxe, L., Gee, J. S., and Pick, T. (1998). The intensity of the Earth's magnetic field over the past 160 million years. *Nature*, 394:878–881.

Katari, K. and Bloxham, J. (2001). Effects of sediment aggregate size on DRM intensity :A new theory. *Earth Planet. Sci. Lett.*, 186:113–122.

King, J. W., Banerjee, S. K., and Marvin, J. A. (1983). A new rock magnetic approach to selecting sediments for geomagnetic paleointensity studies: application to paleointensity for the last 4000 years. *J. Geophys. Res.*, 88:5911–5921.

Kono, M. and Hiroi, O. (1996). Paleosecular variation of field intensites and dipole moments. *Earth Planet. Sci. Lett.*, 139:251–262.

Krása, D., Heunemann, C., Leonhardt, R., and Petersen, N. (2003). Experimental procedure to detect multidomain remanence during thellier-thellier experiments. *Phys. Chem. Earth*, 28:681–687.

Krása, D., Shcherbakov, V. P., Kunzmann, T., and Petersen, N. (2005). Self-reversal of remanent magnetization in basalts due to partially oxidized titanomagnetites. *Geophys. J. Int.*, 162:115–136.

Laj, C., Kissel, C., Mazaud, A., Channell, J., and Beer, J. (2000). North Atlantic paleointensity stack since 75 ka (NAPIS-75) and the duration of the Laschamp event. *Phil. Trans. Roy. Soc.*, 3358:1009–1025.

Laj, C., Mazaud, A., Weeks, R., Fuller, M., and Herrero-Bevera, E. (1991). Geomagnetic reversal paths. *Nature*, 351:447.

Langereis, C., Dekkers, M., de Lange, G., Paterne, M., and van Santvoort, P. (1997). Magnetostratigraphy and astronomical calibration of the last 1.1 Myr from an eastern Mediterranean piston core and dating of short events in the Brunhes. *Geophys. J. Int.*, 129:75–94.

Leonhardt, R. (2006). Analyzing rock magnetic measurements: The Rockmaganalyzer 1.0 software. *Comp. Geosci.*, 32, 1420–1431.

Leonhardt, R. and Fabian, K. (2007). Paleomagnetic reconstruction of the global geomagnetic field evolution during the Matuyama-Brunhes transition: Iterative Bayesian inversion and independent verification. *Earth Planet. Sci. Lett.*, 253:172–195.

Leonhardt, R., Hufenbecher, F., Heider, F., and Soffel, H. C. (2000). High absolute paleointensity during a mid Miocene excursion of the Earth's magnetic field. *Earth Planet. Sci. Lett.*, 184:141–154.

Leonhardt, R., Matzka, J., Hufenbecher, F., Heider, F., and Soffel, H. (2002). A reversal of the Earth's magnetic field recorded in mid Miocene lava flows of Gran Canaria: Paleodirections. *J. Geophys. Res.*, 107(B1):2024, doi:10.1029/2001JB000322.

Leonhardt, R., Matzka, J., and Menor, E. A. (2003). Absolute paleointensities and paleodirections from Fernando de Noronha, Brazil. *Phys. Earth Planet. Inter.*, 139:285–303.

Leonhardt, R., Heunemann, C., and Krása, D. (2004a). Analyzing absolute paleointensity determinations: Acceptance criteria and the software Thelliertool4.0. 5:Q12016, doi:10.1029/2004GC000807.

Leonhardt, R., Krása, D., and Coe, R. S. (2004b). Multidomain behavior during Thellier paleointensity experiments: A phenomenological model. *Phys. Earth Planet. Inter.*, 147:127–140.

Leonhardt, R., Matzka, J., Nichols, A., and Dingwell, D. (2006). Cooling rate correction of paleointensity determination for volcanic glasses by relaxation geospeedometry. *Earth Planet. Sci. Lett.*, 243:282–292.

Leonhardt, R. and Soffel, H. C. (2002). A reversal of the Earth's magnetic field recorded in mid Miocene lava flows of Gran Canaria: Paleointensities. *J. Geophys. Res.*, 107(B11):2299, doi:10.1029/2001JB000949.

Leonhardt, R. and Soffel, H. C. (2006). The growth, collapse and quiescence of Teno volcano, Tenerife: new constraints from paleomagnetic data. *Int. J. Earth Sci. (Geol. Rundsch.)*, 95:1053–1064.

Levi, S. (1977). The effect of magnetite particle size on paleointensity determinations of the geomagnetic field. *Phys. Earth Planet. Inter.*, 13:245–259.

Levi, S. and Banerjee, S. K. (1976). On the possibility of obtaining relative paleointensities from lake sediments. *Earth Planet. Sci. Lett.*, 29:219–226.

Lund, S., Stoner, J., Channell, J., and Acton, G. (2006). A summary of Brunhes paleomagnetic field variability recorded in Ocean Drilling Program cores. *Phys. Earth Planet. Inter.*, 156:194–204.

Martinson, D., Pisias, N., Hays, J., Imbrie, J., Moore, T.C., J., and Shackleton, N. (1987). Age dating and the orbital theory of the ice ages of a high-resolution 0 to 300,000-year chronostratigraphy. *Quat. Res.*, 27:1–29.

Mazaud, A., Sicre, M., Ezat, U., Pichon, J., Duprat, J., Laj, C., Kissel, C., Beaufort, L., Michel, E., and Turon, J. (2002). Geomagnetic-assisted stratigraphy and sea surface temperature changes in core MD94-103 (Southern Indian Ocean): possible implications for North-South climatic relationships around H4. *Earth Planet. Sci. Lett.*, 201:159–170.

McClelland-Brown, E. (1984). Experiments on trm intensity dependence on cooling rate. *Geophys. Res. Lett.*, 11:205–208.

McElhinny, M. W. and Lock, J. (1996). IAGA paleomagnetic databases with access. *Surv. Geophys.*, 17:575–591.

Néel, L. (1949). Théorie du traînage magnétique des ferromagnétiques en grains fins avec applications aux terres cuites. *Ann. Geophys.*, 5:99–136.

Nowaczyk, N. and Baumann, M. (1992). Combined high-resolution magnetostratigraphy and nannofossil biostratigraphy for late Quaternary Arctic Ocean sediments. *Deep Sea Res.*, 39:567–601.

Nowaczyk, N. R. and Knies, J. (2000) Magnetostratigraphic results from Eastern Artic Ocean-AMS14C ages and relative paleointensity data of the Mono Lake and Laschamp geomagnetic events. *Geophys. J. Int.*, 140:185–197.

Nowaczyk. N. R., Antonow, M., Knies, J., Spielhagen, R. F. (2003) Further magnetostratigraphic results on reversal excursions during the last 50 ka derived from northern high latitudes and discrepancies in their precise AMS14C dating. *Geophys. J. Int.*, 155:1065–1080.

Opdyke, N. D. and Channell, J. E. T. (1996). *Magnetic Stratigraphy*. Academic Press, International Geophysics Series, Volume 64, San Diego, USA.

Otofuji, Y. and Sasajima, S. (1981). A magnetization process of sediments: laboratory experiments on post-depositional remanent magnetization. *GeoPhys. J. R. Astr. Soc.*, 66:241–259.

Perrin, M. and Shcherbakov, V. (1997). Paleointensity of the Earth's magnetic field for the past 400 ma: Evidence for a dipole structure during the mesozoic low. *J. Geomag. Geoelectr.*, 49:601–614.

Piper, J. D. A. and Richardson, A. (1972). The palaeomagnetism of the Gulf of Guinea volcanic province, West Africa. *Geophys. J. R. Astr. Soc.*, 29:147–171.

Prévot, M. and Camps, P. (1993). Absence of preferred longitude sectors for poles from volcanic records of geomagnetic reversals. *Nature*, 366:53–57.

Prévot, M., Derder, M., McWilliams, M., and Thompson, J. (1990). Intensity of the Earth's magnetic field: evidence for a Mesozoic dipole low. *Earth Planet. Sci. Lett.*, 97:129–139.

Reid, J. (1989). On the total geostrophic circulation of the South Atlantic Ocean: Flow patterns, tracers, and transports. *Prog. Oceanog.*, 23:149–244.

Richardson, A. and Watkins, N. D. (1967). Paleomagnetism of Atlantic islands: Fernando Noroñha. *Nature*, 215:1470–1473.

Riisager, P. and Riisager, J. (2001). Detecting multidomain magnetic grains in Thellier paleointensity experiments. *Phys. Earth Planet. Inter.*, 125:111–117.

Roberts, A. P. and Winklhofer, M. (2005) Why are geomagnetic excursions not always recorded in sediments? Constraints from post-depositional remanent magnetization lock-in modelling. *Earth Planet. Sci. Lett.*, 227:345–359.

Schmieder, F. (2004). Magnetic signals in Plio-Pleistocene sediments of the South Atlantic: Chronostratigraphic usability and paleoceanographic implication. In Wefer, G., Mulitza, S., and Ratmeyer, V., editors, *The South Atlantic in the Late Quaternary: Reconstruction of Material Budgets and Current Systems*, pp. 269–305. Springer-Verlag, Berlin.

Schult, A., Calvo Rathert, M., Guerreiro, S., and Bloch, W. (1986). Paleomagnetism and rock magnetism of Fernando de Noronha, brazil. *Earth Planet. Sci. Lett.*, 79:208–216.

Selkin, P. A. and Tauxe, L. (2000). Long-term variations in paleointensity. *Philos. Trans. R. Soc.*, 358:1065–1088.

Shaw, J. (1974). A new method of determining the magnetide of the paleomagnetic field: Application to five historic lavas and five archeological samples. *Geophys. J. R. Astr. Soc.*, 39:133–141.

Shcherbakov, V. P. and Shcherbakova, V. V. (1987). On the physics of post-depositional remanent magnetization. *Phys. Earth Planet. Inter.*, 46:64–70.

Shcherbakov, V. P., Solodovnikov, G. M., and Sycheva, N. K. (2002). Variations in the geomagnetic dipole during the past 400 million years (volcanic rocks). *Phys. Solid Earth,* 38:113–119.

Stacey, F. D. (1972). On the role of Brownian motion in the control of detrital remanent magnetization in sediments. *Pure Appl. Geophys.*, 98:139–145. IRM library.

Stoner, J., Channell, J., Hodell, D., and Charles, C. (2003). A ∼580 kyr paleomagnetic record from the sub-Antarctic South Atlantic (Ocean drilling Program Site 1089). *J. Geophys. Res.*, 108:doi:10.1029/2001JB001390.

Stoner, J., Laj, C., Channell, J., and Kissel, C. (2002). South Atlantic and North Atlantic geomagnetic paleointensity stacks (0- 80 ka): implications for inter-hemispheric correlation. *Quat. Sci. Rev.*, 21:1141–1151.

Stramma, L. and England, M. (1999). On the water masses and mean circulation of the South Atlantic Ocean. *J. Geophys. Res.*, 104(C9):20863–20883.

Stramma, L. and Peterson, R. (1990). The South Atlantic current. *J. Phys. Ocean.*, 20:846–859.

Talley, L. (1996). Antarctic intermediate water in the South Atlantic. In Wefer, G., Berger, W., Siedler, G., and Webb, D., editors, *The South Atlantic: Present and Past Circulation* , pp. 219–238. Springer Verlag, Berlin.

Tanaka, H., Kono, M., and Uchimura, H. (1995). Some global features of palaeointensity in geological time. *Geophys. J. Int.*, 120:883–896.

Tarduno, J. A. and Smirnov, A. V. (2001). Stability of the earth with respect to the spin axis for the last 130 million years. *Earth Planet. Sci. Lett.*, 184:549–553.

Tauxe, L. (1993). Sedimentary records of relative paleointensity: Theory and practice. *Rev. Geophys.*, 31:319–354.

Thellier, E. (1941). Sur la vérification d'une méthode permettant de déterminer l'intensité du champ magnétique terrestre dans le passé. *C. R. Acad. Sci.*, 212:281–283.

Thellier, E. and Thellier, O. (1959). Sur l'intensité du champ magnétique terrestre dans le passé historique et géologique. *Annales de Géophysique*, 15:285–376.

Theyer, F., Herrero-Bervera, E., and Hsu, V. (1985). The zonal harmonic model of polarity transitions: a test using successive reversals. *J. Geophys. Res.*, 90:1963–1982.

Thouveny, N., Carcaillet, J., Moreno, E., Leduc, G., and Nerini, D. (2004). Geomagnetic moment variation and paleomagnetic excursions since 4000 kyr BP: a stacked record from sedimentary sequences of the Portuguese margin. *Earth Planet. Sci. Lett*, 219:377–396.

Thouveny, N., Creer, K., and Blunk, I. (1990). Extension of the Lac du Bouchet palaeomagnetic record over the last 120000 years. *Earth Planet. Sci. Lett.*, 97:140–161.

Thouveny, N., Creer, K., and Williamson, D. (1993). Geomagnetic moment variations in the last 70000 years, impact on production of cosmogenic isotopes. *Glob. Planet. Change*, 7:157–172.

Tucker, P. (1980). Stirred remanent magnetization: a laboratory analogue of post-depositional realignment. *J. Geophys.*, 48:153–157.

Tucker, P. (1981a). Low-temperature magnetic hysteresis properties of multidomain single-crystal titanomagnetite. *Earth Planet. Sci. Lett.*, 54:167 – 172.

Tucker, P. (1981b). Paleointensities from sediments: normalization by laboratory redepositions. *Earth Planet. Sci. Lett.*, 56:398–404.

Valet, J.-P., Brassat, J., Quidelleur, X., Soler, V., Gillot, P.-Y., and Hongre, L. (1999). Paleointensity variations across the last geomagnetic reversal at La Palma, Canary Islands, Spain. *J. Geophys. Res.*, 104:7577–7598.

Valet, J.-P. and Herrero-Bervera, E. (2003). Some characteristics of geomagnetic reversals inferred from detailed volcanic records. *Compt. Ren. Geosci.*, 335:79–90.

Valet, J.-P., Tucholka, P., Courtillot, V. E., and Meynadier, L. (1992). Palaeomagnetic constraints on the geometry of the geomagnetic field during reversals. *Nature*, 356:400–407.

van Vreumingen, M. J. (1993a). The influence of salinity and flocculation upon the acquisition of remanent magnetization in some artificial sediments. *Geophys. J. Int.*, 114:607–614.

van Vreumingen, M. J. (1993b). The magnetization intensity of some artificial suspensions while flocculating in a magnetic field. *Geophys. J. Int.*, 114:601–606.

Veitch, R. J., Hedley, I. G., and Wagner, J.-J. (1984). An investigation of the intensity of the geomagnetic field during Roman times using magnetically anisotropic bricks and tiles. *Arch. Sc. Genéve*, 37:359–373.

von Dobeneck, T. and Schmieder, F. (1999). Using rock magnetic proxy records for orbital tuning and extended time series analyses into the super- and sub-Milankovitch Bands. In Fischer, G. and Wefer, G., editors, *Use of Proxies in Paleoceanography: Examples from the South Atlantic*, pp. 601–633. Springer-Verlag, Berlin.

Wefer, G. and cruise participants (2001). Report and preliminary results of Meteor Cruise M46/4, Mar del Plata (Argentinia)-Salvador da Bahia (Brazil), February 10- March 13, 2000. With partial results of Meteor cruise M46/2. *Berichte, FB Geowissenschaften, Universität Bremen*, 173.

Wehland, F., Leonhardt, R., Vadeboin, F., and Appel, E. (2005) Magnetic interaction analysis of basaltic samples and preselection for absolute paleointensity measurements. *Geophys. J. Int.*, 162:315–320.

Williams, I. and Fuller, M. (1981). Zonal harmonic models of reversal transition fields. *J. Geophys. Res.*, 86(B12):11657–11665.

# Chapter 4
# Numerical Models of the Geodynamo: From Fundamental Cartesian Models to 3D Simulations of Field Reversals

Johannes Wicht, Stephan Stellmach and Helmut Harder

## 4.1 Formulating the Dynamo Problem

Numerical dynamo simulations by Glatzmaier and Roberts (1995) mark a break-through in dynamo physics. Earlier computer simulations modeling the generation of magnetic field by convectively driven flows had proven the general validity of the concept (Zhang and Busse, 1988, 1989, 1990). The work by Glatzmaier and Roberts (1995), however, is regarded as the first realistic simulation of the geodynamo process. Most remarkably, they presented a magnetic field reversal that resembles many features observed in paleomagnetic data.

A growing number of numerical models explored various aspects of the dynamo process since then. Most approaches were geared to explain the geomagnetic field and adopted a spherical shell geometry, some other models tackled more fundamental questions in a simpler box geometry that offers numerical benefits (St. Pierre, 1993; Stellmach and Hansen, 2004). In a more recent development, numerical simulations also set out to explain the diverse internal magnetic fields of the other planets in our solar system (Stanley and Bloxham, 2004; Stanley et al., 2005; Takahashi and Matsushima, 2006; Christensen, 2006; Glassmeier et al., 2007).

Numerical dynamos not only explain strength and large scale geometry of planetary fields, but also model many of the observed details and variations. They replicate, for example, the gross location of inverse and normal magnetic flux patches at Earth's core–mantle boundary and also help to understand their origin. This broad success is somewhat surprising, since numerical limitations force dynamo modelers to run their simulations at parameters that are far away from realistic values. In particular, the fluid viscosity is generally chosen many orders of magnitude too large in order to damp the small scale turbulent structures that can not be resolved

Johannes Wicht

Max-Planck-Institut für Sonnensystemforschung Max-Planck-Straße 2 37191 Katlenburg-Lindau Germany

Stephan Stellmach

Earth Sciences Department University of California Santa Cruz, CA 95064 USA

Helmut Harder

Institut für Geophysik Universität Münster Corrensstrasse 24 48149 Münster Germany

K.-H. Glaßmeier et al. (eds.), *Geomagnetic Field Variations*, Advances in Geophysical and Environmental Mechanics and Mathematics,
© Springer-Verlag Berlin Heidelberg 2009

numerically. Recent scaling analyses, however, suggest that the success may not be coincidental, after all, since geomagnetic and numerical dynamos may indeed work in the same regime (Christensen and Tilgner, 2004; Christensen and Aubert, 2006; Olson and Christensen, 2006). Though today's simulations cannot resolve the small scale turbulence, they nevertheless seem to model the larger scale dynamo process quite realistically. A concise overview of the achievements and failures of modern dynamo modeling is given by Christensen and Wicht (2007).

Here, we concentrate on three main topics. After a short introduction into the dynamo problem, we proceed with our first main subject, the advancement of numerical methods. Though the present simulations are already quite successful, we also aim at running dynamo simulations at more and more realistic parameters. The numerical benefits of local numerical methods, that abandon the more classical pseudo-spectral approach, may help here. In Sect. 4.2 we summarize recent advances in applying local methods to dynamo simulations. In Sect. 4.3 we discuss simulations in a cartesian box system, that allows us to better isolate and investigate some fundamental problems and ideas than in the more complex spherical shell geometry. Finally, we illustrate the usefulness of dynamo simulations for interpreting and understanding geomagnetic processes in Sect. 4.4, where we explain how the fundamental dipole dominated geomagnetic field geometry is established and why and how it breaks down during excursions and reversals.

What we call the dynamo process refers to the creation of magnetic field by electromagnetic induction: When an electrical conductor moves through an already present initial magnetic field, electric currents are excited that in turn establish their own magnetic field. The electrical conductivity of Earth's iron core and the convective motions in its outer liquid part provide two basic ingredients for a working dynamo. The spherical shell bounded by the solid inner core and the core–mantle boundary forms Earth's active dynamo region, where convective motions are driven by two effects in similar proportions (Lister and Buffett, 1995): the heat flux out of the shell and the release of light elements from the solidifying inner core.

A third issue to consider in a dynamo process is the initial field. In a self excited dynamo, the newly created field can take the role of the initial field, i.e. no additional field is required. Most planetary and lunar dynamos likely fall into this category with the exception of Io, possibly also Ganymede and Mercury. When the initial magnetic field can be infinitely small, we talk about a supercritical dynamo where small magnetic disturbances can grow to a full fledged dynamo field. Again, this seems to be the most likely case for self excited planetary dynamos.

The rotation of the spherical shell plays an important role in organizing the flow and thereby the magnetic field and is a necessary ingredient to receive a magnetic field that is dominated by the axial dipole component (see Sect. 4.4.1). The fast rotation guarantees that the flow dynamics approaches the so-called magnetostrophic regime, where Coriolis force, Lorentz force, and pressure gradient determine the leading-order flow structure. We will explore the force balance in more detail in section Sect. 4.3.

Figure 4.1 shows the basic setup adopted in a spherical shell dynamo. The simulations solve for fluid flow $\mathbf{U}$, magnetic field $\mathbf{B}$, pressure $p$, and density differences

**Fig. 4.1** Setup of dynamo model

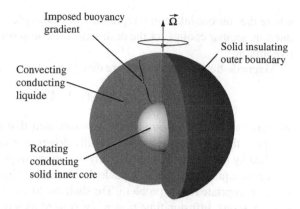

Imposed buoyancy gradient

Solid insulating outer boundary

Convecting conducting liquide

Rotating conducting solid inner core

$\delta\rho = \alpha T$ in the spherical shell. Density differences can have a thermal and/or a compositional origin. The parameter $\alpha$ can therefore be interpreted as the thermal expansivity or a compositional equivalent, depending on whether $T$ stands for super-adiabatic temperature perturbations or local variation in the amount of light constituents (Lister and Buffett, 1995; Wicht et al., 2007). Most simulations do not distinguish between the two driving types, assuming that both possess similar turbulent diffusivities and thus obey the same advection/diffusion equation (Kutzner and Christensen, 2000). A notable exception is the work by Glatzmaier and Roberts (1996) where thermal and compositional density variations are treated separately. However, a conclusive exploration of the possible implication of double-diffusive convection is still missing. For simplicity, we will mostly refer to $T$ as the (super-adiabatic) temperature in the following.

Flow changes are described by the Navier–Stokes equation:

$$E\left(\frac{\partial \mathbf{U}}{\partial t} + \mathbf{U} \cdot \nabla \mathbf{U}\right) + 2\hat{\mathbf{z}} \times \mathbf{U} + \nabla \Pi =$$
$$E \nabla^2 \mathbf{U} + Ra\, Pr^{-1} \frac{\mathbf{r}}{r_o} T + \frac{1}{Pm} (\nabla \times \mathbf{B}) \times \mathbf{B}, \qquad (4.1)$$

where unit vector $\hat{\mathbf{z}}$ points in the direction of the rotation axis. The modified pressure $\Pi$ combines the non–hydrostatic pressure and centrifugal forces.

The induction equation, derived from Maxwell's laws and Ohm's law, determines the magnetic field evolution:

$$\frac{\partial \mathbf{B}}{\partial t} - \nabla \times (\mathbf{U} \times \mathbf{B}) = \frac{1}{Pm} \nabla^2 \mathbf{B}. \qquad (4.2)$$

The evolution of the super-adiabatic temperature follows the transport equation

$$\frac{\partial T}{\partial t} + \mathbf{U} \cdot \nabla T = \frac{1}{Pr} \nabla^2 T + \varepsilon, \qquad (4.3)$$

where the source/sink term $\varepsilon$ represents, for example, possible radiogenic heat production, secular cooling, or the destruction of compositional differences by convective mixing.

Magnetic field and flow field are divergence free:

$$\nabla \cdot \mathbf{B} = 0, \ \nabla \cdot \mathbf{U} = 0 .$$

(4.4)

We have adopted the Boussinesq approximation that neglects all density changes except in the buoyancy term that actually drives convection. This approximation is justified by the relatively small density variations in planetary iron cores.

Above equations have been made dimensionless by adopting a set of scales that seem appropriate for the problem. The shell thickness $d = r_o - r_i$ serves as the length scale, viscous diffusion time $t_v = d^2/v$ is used as a time scale, the magnetic scale is $(\rho\mu\lambda\Omega)^{1/2}$, and the temperature difference $\Delta T$ across the shell serves as the temperature scale. Here, $r_o$ and $r_i$ are outer and inner boundary radii, respectively, $v$ is the kinematic viscosity, $\rho$ the mean fluid density, $\mu$ the magnetic permeability, $\lambda$ the magnetic diffusivity, $\kappa$ the thermal or chemical diffusivity, and $\Omega$ the shell rotation rate.

Four dimensionless parameters appear in the set of equations:
Rayleigh number

$$Ra = \frac{\alpha g_o \Delta T d}{\kappa \Omega} ,$$

(4.5)

Ekman number

$$E = \frac{v}{\Omega d^2} ,$$

(4.6)

Prandtl number

$$Pr = \frac{v}{\kappa} ,$$

(4.7)

magnetic Prandtl number

$$Pm = \frac{v}{\lambda} .$$

(4.8)

We have used the gravity acceleration $g_o$ at the outer boundary as a reference and assume that gravity changes linearly with radius. The Rayleigh number differs from the "classical" definition in that, for example, rotational effects are included (Kono and Roberts, 2001). Different authors choose different set of scales and dimensionless parameters, we largely follow the definitions established for the benchmark dynamo (Christensen et al., 2001).

Typically, rigid flow boundary conditions imply that radial and latitudinal flow components vanish at $r = r_i$ and $r = r_o$ and that the azimuthal flow components have to match the rotation of inner core and/or mantle, respectively. Since the electrical conductivity of the mantle is many orders of magnitude lower than that of the core, it can be regarded as an insulator in comparison. The core field then has to match a potential field at $r = r_o$, i.e it has to obey

$$\nabla^2 B = 0 \quad \text{for} \quad r \geq r_o. \tag{4.9}$$

Most dynamo models assume an electrically conducting inner core, where a modified induction Eq. (4.2) is solved, replacing flow U with the inner core solid body rotation. Appropriate matching conditions guarantee the continuity of the magnetic field and the horizontal electric field at the inner core boundary (Wicht, 2002; Christensen and Wicht, 2007). The inner core can rotate freely about the $z$-axis in response to viscous and Lorentz torques. Rotations about any axis in the equatorial plane are generally neglected since they would be damped by strong gravitational torques due to the inner-core and mantle oblatenesses. Viscous and Lorentz torques also act on the mantle, but the associated changes in rotation rate are very small due to the mantle's large moment of inertia and can be neglected in the dynamo context. For more details on the system of equations and different model approaches see, for example, Christensen and Wicht (2007).

A few diagnostic parameters are handy for analyzing and comparing different dynamos. Reynolds number $Re$ measures the ratio of inertial to viscous effects in the Navier–Stokes equation (4.1) and is identical to the RMS flow amplitude $U$ in the scaling used here:

$$Re = U = \sqrt{\langle \mathbf{U}^2 \rangle}, \tag{4.10}$$

triangular brackets denote the mean over the spherical shell. The magnetic Reynolds number $Rm$ measures the ratio of induction and diffusion terms in the dynamo Eq. (4.2):

$$Rm = Ud/(\lambda). \tag{4.11}$$

The Elsasser number $\Lambda$ measures the importance of the Lorentz force in the Navier–Stokes equation, more precisely the ratio of Lorentz to Coriolis force. Note that $\Lambda$ is also a measure for the RMS magnetic field strength $B$ in our scaling:

$$\Lambda = B^2/2. \tag{4.12}$$

The magnetic field on a spherical surface is typically decomposed into spherical harmonic contributions of order $l$ and degree $m$. This allows to calculate the energy $\Lambda(l,m)$ carried by a single mode. The Mauersberger-Lowes spectrum (Mauersberger, 1956), often used to characterize the magnetic field at Earth's surface or the core mantle boundary, quantifies the magnetic energy for each degree $l$: $\Lambda(l) = \sum_m \Lambda(l,m)$. A similar decomposition of the flow field allows to derive respective kinetic energy contributions $E_k(l,m)$ and $E_k(l)$. We use the latter to calculate the typical latitudinal flow length scale $\ell$ that has been introduced by Christensen and Aubert (2006):

$$\ell = d\pi/\bar{l}, \tag{4.13}$$

with

$$\bar{l} = \frac{\sum_l l E_k(l)}{\sum_l E_k(l)}. \tag{4.14}$$

In order to quantitatively compare the numerical runs to the geomagnetic field we have re-scaled time with the outer core magnetic diffusion time, which amounts to 126 kyr when assuming the following properties: $r_o = 3,485$ km, $r_i = 1,222$ km, and $\sigma = 6 \times 10^5$ S m$^{-1}$. Note, however, that this is not the only way to re-scale time. For re-scaling the magnetic field strength we had to quantify two additional properties: the mean inner core density, $\rho = 11 \times 10^3$ kg m$^{-3}$, and Earth's rotation rate, $\Omega = 7.29 \times 10^{-5}$ sec$^{-1}$.

## 4.2 Numerical Methods

### 4.2.1 Spectral Versus Local Approaches

Today, spectral approaches are the most established method for numerical dynamo simulations. First applied in the dynamo context by Bullard and Gellman (1954), they have been optimized for the spherical shell dynamo problem over the years. Notably, Glatzmaier (1984) has established a numerical scheme that many authors have followed since.

The core of this pseudo-spectral approach is the dual representation of each variable in grid space as well as in spectral space. The individual computational steps are performed in whichever representations they are most efficient. Spherical harmonic functions $Y_{lm}$ are the obvious choice for expanding latitudinal and longitudinal dependencies. They also have the advantage of being Eigen-solutions of the Laplace operator, i.e. they are the natural choice for a diffusive problem. Chebychev polynomials $C_n$ are typically chosen for radial representations, where $n$ is the degree of the polynomial. When applied appropriately, they offer the advantage of a higher resolution near the inner and outer boundaries, where boundary layers may have to be sampled. They also allow to employ a Fast Fourier transform for switching between grid to spectral representations (Glatzmaier, 1984).

The differential equation system is time-stepped in spherical harmonic-radial space $(l, m, r)$ using a mixed implicit/explicit scheme that handles the non-linear terms as well as the Coriolis force in an explicit Adams-Bashforth time step. This guarantees that all spherical harmonic modes $(l, m)$ decouple. A Crank-Nicolson implicit scheme completes the time integration for the remaining terms.

The nonlinear terms are evaluated in local grid space, which requires a transform from $(l, m, r)$ to $(\phi, \theta, r)$ representation and back ($\phi$ and $\theta$ denote longitude and colatitude respectively). Fast Fourier transforms can be applied for the longitudinal dependence. The Gauss-Legendre transforms, employed in latitudinal direction, are significantly more time consuming and can considerably slow down the computation for highly resolved cases. Generally, all partial derivatives are evaluated in spectral space, which is the only raison d'être for the Chebychev representation in radius. This approach guarantees a high degree of exactness that other radial approximations only reach at the cost of a considerably denser numerical grid (Christensen et al., 2001). The dynamo models by Kuang and Bloxham (1999) and Dormy et al. (1998) apply finite differences in radial direction but retain the spherical harmonic representation, thus forming an intermediate step between fully spectral and fully local methods that we present in the following. More information on the spectral method described here can be found in Glatzmaier (1984) and Christensen and Wicht (2007).

The advantages of spectral methods are most obvious at high or moderate Ekman numbers where the diffusion terms are sizeable contributions in the force balance, where smooth solutions therefore prevail, and a few harmonics suffice to capture most of the energy. The exponential convergence of a spectral method ensures extremely accurate solutions in this regime, which has been demonstrated, for example, by the dynamo benchmark study of Christensen et al. (2001). Similar conclusions can be derived from the benchmark results for the much simpler case of two-dimensional Cartesian convection with constant viscosity (Blankenbach et al., 1989). This early two-dimensional study also demonstrated that the situation is reversed in favor of local methods for more complex high Rayleigh number problems. While this result can possibly not be generalized it, nevertheless, indicates that local methods may become more competitive when the parameters approach more realistic values. Reaching smaller Ekman numbers is a major aim of present dynamo simulations, which is a demanding challenge since the solutions become increasingly small-scaled for decreasing Ekman numbers and are characterized by almost flat spectra.

In this context, the interest in applying local methods to dynamo problems has much increased in recent years, whereas in the past spectral methods were used almost exclusively, the compressible finite difference approach of Kageyama and coworkers (Kageyama et al., 1993, 1997) being a rare exception. Local methods, like finite elements, finite volume, or finite difference approaches, offer some important advantages compared to spectral methods. There are much more choices in structuring the grid, implicit time stepping is easier to implement, local variations of material properties like viscosity or electric conductivity are easier to cope with, and the discretisation of the non-linear terms (advection and Coriolis force) is more flexible than in spectral approaches.

The most cited reason in favor of local methods is, however, the expected better performance on massively parallel computing systems. Low Ekman number computations for $E < 10^{-5}$ are only manageable with parallel computations. Here, local methods have the particular advantage that, with the use of a suitable domain decomposition, only next neighbor communication is required, and only two-dimensional data structures at the domain boundaries must be exchanged. Spectral methods, however, require global communication, i.e. the entire solution must be exchanged during a single time step. Therefore, the cross-processor communication is less demanding for local methods. Naturally, efficient parallel implementations are also possible for spectral methods; Clune et al. (1999) have demonstrated this. However, if the problem size and number of used processes exceed a certain limit, local methods should become more efficient. Unfortunately, it is still unknown whether this expected takeover materializes at realistic conditions. At present, we can only speculate which is the method of choice for dynamo simulations at low Ekman number.

## *4.2.2  Recent Implementations of Local Methods*

In the following, we give a short description of recent implementations of local methods for the simulation of hydromagnetic dynamos in spherical shells. We restrict the compilation to approaches where published comparisons to at least one of the benchmark cases defined in Christensen et al. (2001) are available. For completeness, the recent Yin-Yang grid approach of Kageyama and Yoshida (2005) must also be mentioned although no benchmark comparison is available yet. This finite difference method utilizes an overset grid consisting of two overlapping longitude-latitude grids and two perpendicular frames of reference. This avoids the polar problem of spherical coordinates in an elegant manner. Kageyama and Yoshida (2005) report dynamo simulations with up to nearly $10^9$ gridpoints using up to 4096 parallel processes on the Earth Simulator system, which demonstrates the possibilities of parallel computing. We review a variety of different local approaches which, nevertheless, also share some common features. Notably, primitive variables are used throughout, i.e. the pressure is retained in the solution. A divergence free velocity field is enforced iteratively by solving a Poisson equation for the pressure or pressure correction, but the detailed strategies are different in each approach. This is in sharp contrast to spectral approaches where the spectral base functions for the flow are constructed in such way that the continuity equation is fulfilled identically, in most implementations by the use of poloidal and toroidal stream functions. The reason for these different strategies is simply that the lateral Laplace operator acting on a spherical harmonic expansion is trivial to compute which simplifies a spectral stream function solution. In local approaches a stream function formulation would require large computational stencils which are cumbersome to evaluate. A primitive variable formulation is therefore much more popular in local approaches.

### 4.2.2.1  Harder and Hansen

A finite volume method was developed by Harder and Hansen (2005). In this approach, the lateral grid on a spherical surface is created by projecting a grid from the surfaces of an inscribed cube to the unit sphere. The grid is nearly orthogonalized by successive averaging positions over neighboring grid points. In the radial direction, the grid is usually refined towards the boundaries by a smooth stretching function. Equations and vector quantities (velocity and magnetic field vector) are formulated with reference to a Cartesian coordinate system. A collocated arrangement of variables is preferred instead of the usual staggered approach. A pressure weighted interpolation scheme (PWI) of the normal velocity component towards the cell surfaces avoids the pressure decoupling problem. The conditions $\nabla \cdot \mathbf{U} = 0$ and $\nabla \cdot \mathbf{B} = 0$ are fulfilled by an auxiliary potential approach. Time stepping is performed by a backward, three level finite difference operator of second order accuracy where the Coriolis force is incorporated implicitly. The discrete Navier–Stokes and magnetic induction equations are solved by a block-Jacoby iteration, whereas Poisson equations for corrections of the pressure and magnetic pseudo-pressure are solved by a preconditioned conjugate gradient method. Usually, the quasi-vacuum condition is applied as the magnetic boundary condition, setting the tangential components to $B_{\mathrm{tang}} = 0$ at the boundary. Recently, insulating magnetic boundaries have also been incorporated by matching a spherical harmonic expansion of the exterior field to the internal field.

### 4.2.2.2 Matsui and Okuda

Matsui and Okuda (2005) reported results for both the non-magnetic benchmark case 0 and the dynamo benchmark case 1 with insulating boundaries. The method is based on the parallel GeoFEM thermal-hydraulic subsystem developed for the Japanese Earth Simulator project. A finite element (FE) approximation is applied utilizing hexahedral elements. Within each element all physical variables are approximated by tri-linear functions. A fractional time stepping method is used with a Crank-Nicolson scheme for the diffusion terms and second order Adams-Bashforth extrapolation for the other terms. The magnetic field $\mathbf{B}$ is represented by a vector potential $\mathbf{B} = \nabla \times \mathbf{A}$. The Coulomb gauge condition and the continuity condition of the velocity field are enforced by the solution of Poisson equations. The conjugate gradient method is used to solve the discrete equations. In the case of insulating magnetic boundary conditions a solution of $\nabla^2 \mathbf{A} = \mathbf{0}$ is calculated outside of the conducting fluid. At sufficient distance $\mathbf{A}$ is set to zero. Different lateral mesh patterns are used for the two cases. In the non-magnetic case 0 an icosahedral triangulation of the spherical surface is used, whereas in the magnetic case 1 a projection of a cube to the sphere is preferred. This is motivated by the need of solving for the magnetic field outside the fluid shell and by the difficulty to fill the center ($\mathbf{r} = \mathbf{0}$) with an icosahedral mesh.

### 4.2.2.3 Fournier and Coauthors

The spectral element method of Fournier et al. (2005) can be regarded as a mixture of local and spectral methods. The method is based on a Fourier expansion of the physical variables (temperature, pressure and velocity) in azimuthal direction. For each Fourier mode the associated problem is solved in the two-dimensional meridional plane by a spectral element approach. In each element the solution is approximated by polynomials of order $N$ and of reduced order $N - 2$ for the pressure. In the results presented by Fournier et al. (2005) $N$ varies between 14 and 22. In most cases the meridional plane is treated as a single macro-element (see Table 4.1) motivated by the smoothness of the benchmark solution. The method is formulated in cylindrical coordinates. This frame of reference is favored by the Proudman-Taylor theorem. This theorem states that the motion in a rapidly rotating fluid is parallel to a plane perpendicular to the rotation axis. Gauss-Lobatto-Legendre quadrature is used in the meridional elements away from the axis of the symmetry. Due to the coordinate singularity at the axis of symmetry, in the elements bordering the axis of

**Table 4.1** Results for the non-magnetic benchmark case 0. The label *mat* refers to Matsui and Okuda (2005), *hey* to Hejda and Reshetnyak (2004), *har* to Harder and Hansen (2005), *cha1* to finite element and *cha2* to finite difference solutions of Chan et al. (2006), *fou* to Fournier et al. (2005), and *ben* to the best benchmark estimate given by Christensen et al. (2001)

| Ref. | $n_r$ | $n_{lat}$ | $E_{kin}$ | $u_\phi$ | $T$ | $\omega$ |
|------|-------|-----------|-----------|----------|-----|----------|
| *mat* | 34 | 1920 | 62.157 | −10.429 | 0.43126 | 0.09446 |
|      | 66 | 7680 | 59.718 | −10.229 | 0.43896 | 0.06198 |
|      | 96 | 30720 | 59.129 | −10.190 | 0.42838 | 0.05527 |
| *hej* | 45 | 45, 64 | 60.825 | −9.978 | 0.4316 | 0.1123 |
|      | 65 | 65, 64 | 60.450 | −10.055 | 0.4312 | 0.0984 |
|      | 85 | 85, 64 | 60.282 | −10.082 | 0.4311 | 0.1338 |
|      | 85 | 85, 96 | 59.414 | −10.175 | 0.4291 | 0.1187 |
| *har* | 24 | 3456 | 58.167 | −10.534 | 0.4335 | 0.7326 |
|      | 32 | 6144 | 58.359 | −10.387 | 0.4313 | 0.4729 |
|      | 48 | 13824 | 58.374 | −10.260 | 0.4295 | 0.3038 |
|      | 64 | 24576 | 58.357 | −10.214 | 0.4289 | 0.2479 |
| *cha1* | 21 | 2562 | 56.986 | −10.320 | 0.4273 | 0.1604 |
|      | 41 | 10242 | 57.884 | −10.160 | 0.4273 | 0.1709 |
| *cha2* | 50 | 80, 80 | 58.994 | −10.496 | 0.4290 | 0.4804 |
|      | 75 | 120, 120 | 58.896 | −10.363 | 0.4301 | 0.3518 |
| *fou* | 19.3 | 14, 31 | 58.288 | −10.153 | 0.4281 | 0.14676 |
|      | 22.6 | 18, 31 | 58.352 | −10.158 | 0.4281 | 0.18127 |
|      | 25.7 | 22, 31 | 58.347 | −10.157 | 0.4281 | 0.18235 |
|      | 47.6 | $4 \times 18, 63$ | 58.347 | −10.157 | 0.4281 | 0.18232 |
| *ben* | | | 58.348 | −10.1571 | 0.42812 | 0.1824 |

symmetry Gauss-Lobatto-Jacoby is preferred where the cylindrical radius is incorporated in the weight function. The method is advanced in time by a second-order accurate BDF-scheme. Discrete equations are solved by a preconditioned conjugate gradient method.

### 4.2.2.4 Heyda and Reshetnyak

Hejda and Reshetnyak (2004) present solutions for the non-magnetic benchmark case 0 obtained by their control volume approach. Hejda and Reshetnyak (2003) give a detailed description of this method. The method utilizes a staggered arrangement of variables and a spherical longitude-latitude grid. The control volume approach simplifies the numerical handling of the polar problem since the surfaces of the faces at the axis ($\Theta = 0$) are zero. Only when the coupling of variables requires values at the pole axis an extrapolation from adjacent cells is performed. The continuity condition of the velocity field is iteratively fulfilled by a pressure correction scheme. The method includes the solution for the magnetic field. Remarkably, here the radial component $B_r$ is not calculated by the magnetic induction equations, but by the $\nabla \cdot \mathbf{B} = 0$ condition. The linear system of equations is solved by a tridiagonal band solver in radial direction and by under-relaxed Gauss-Seidel iteration in lateral directions.

#### 4.2.2.5 Chan, Li and Liao

Chan et al. (2006) describe both a finite element and a finite difference approach. The lateral grid of the FE method is generated by an icosahedral triangulation of a spherical surface. By stacking the lateral triangulation in radial direction a set of pentahedra is created where each pentahedron is in turn divided into four tetrahedra. The tetrahedra are treated as Hood-Taylor elements where velocity and temperature are approximated by piecewise quadratic polynomials whereas the pressure is approximated by piecewise linear polynomials. Using a BDF-2 scheme for the time discretisation, the linearized discrete equations are solved by a BiCGstab iterative solver.

The finite difference method is based on a uniform longitude-latitude grid combined with a non-uniform radial discretisation which is refined towards the boundaries. A staggered arrangement of variables is used where the pressure node is located at the grid cell center, temperature at the corners of the cell, and the velocity components at the centers of the cell surfaces. In order to avoid the pole singularities backward difference formula are employed at points adjacent to the grid poles. An approximative factorization scheme of the Navier–Stokes equations is used to decouple the calculation of temperature and velocity from the pressure which is determined by a Poisson equation. The time stepping follows a Crank-Nicolson scheme. The linearized equations are solved by the parallel iterative library AZTEC.

### 4.2.3 Testing Local Methods

All summarized approaches have been tested for at least one of the benchmark problems given by Christensen et al. (2001). The benchmark uses the typical setup described in Sect. 4.1, the magnetic field solution as well as the parameters and properties of benchmark II are described in Sect. 4.4.1. The Rayleigh numbers of benchmark cases 0 and I are 10% lower than for benchmark II. Another difference is that benchmark II assumes a conducting inner core. Requested quantities of the computed solutions are kinetic energy $E_k$, magnetic energy $E_m = \Lambda$, azimuthal drift frequency $\omega$, and local azimuthal velocity $u_\phi$, local temperature and local magnetic field component $B_\theta$ at a specified position within the solution.

Selected results of the cited local approaches for the nonmagnetic benchmark case are given in Table 4.1. For all approaches solutions at different grid resolutions are available. Here, $n_r$ is the number of grid points in radial direction, and $n_{lat}$ gives the lateral resolution. If separate values for the grid resolution in both lateral direction are available, both numbers are given. Due to the very different approach of Fournier et al. (2005) $n_r$ and $n_{lat}$ have a different definition: $n_r$ is the mean grid resolution, i.e. the third root of the total degrees of freedom, and the two values given in the $n_{lat}$ column give the maximal order of polynomials in the meridional plane and the maximal wave number in the harmonic expansion in azimuthal direction. In most cases, the meridional plane is represented by a single spectral element. Only in the last solution the meridional plane is divided into four elements.

For all quantities, except the drift frequency $\omega$, the results of all methods seem to converge with increasing grid resolution towards the best spectral benchmark estimate. Unfortunately, the resolution is not always increased homogenously in all spatial directions which can introduce some bias in any kind of data extrapolation. The lack of consistent grid refinement also makes it difficult to verify the convergence of the individual methods. However, Harder and Hansen (2005) have demonstrated that with consistent grid refinement and using Romberg extrapolation very precise estimates can be derived for all requested benchmark data from solutions with successively refined grids. Considering only single grid results, the spectral element approach of Fournier et al. (2005) is certainly the most accurate method, since it shares the exponential convergence rate with the more tractional spectral approaches.

The results for the drift frequency are less accurate than for the other quantities. This was also observed in the benchmark study of Christensen et al. (2001) where various variants of spectral and semi-spectral methods were compared. However, here the problem is more severe since the convergence to the correct benchmark estimate is not obvious for some methods. Certainly, the drift frequency is a very sensible quantity to compute. As pointed out by Matsui and Okuda (2005), the drift frequency of the benchmark case is a very small quantity which may be hard to resolve numerically. Also, as observed by Harder and Hansen (2005), the benchmark case is close to the bifurcation to a time-dependent solution, so that a small numerical error can induce a large variation in the solution. Other causes may also add to the problem: too long time steps or an insufficient reduction of the residuals by the iterative solvers. A conclusive answer is not available at the moment.

For the dynamo simulation with insulating boundaries (benchmark case I) only a few solutions are available (see Table 4.2), notably those obtained by Matsui and Okuda (2005) and some preliminary, unpublished simulations calculated with the approach by Harder and Hansen (2005). The main difficulty in solving this case with a local approach is not the addition of the magnetic induction equations, but the implementation of insulating boundary conditions for the magnetic field, which requires either a solution of the magnetic field in the exterior region or the matching of series expansions at the boundaries. Matsui and Okuda (2005) use the former approach, whereas Harder and Hansen (2005) prefer the latter. As in the previous

**Table 4.2** Results for the magnetic benchmark case 1 with insulating boundaries. Labels as before

| Ref. | $n_r$ | $n_{lat}$ | $E_{kin}$ | $E_{mag}$ | $u_\phi$ | $T$ | $B_\theta$ | $\omega$ |
|------|-------|-----------|-----------|-----------|----------|-----|------------|----------|
| *mat* | 24 | 1944 | 34.455 | 663.89 | −6.7945 | 0.361678 | −5.0835 | 3.1792 |
|       | 24 | 3456 | 33.266 | 643.37 | −7.4634 | 0.372633 | −5.0262 | 3.2223 |
|       | 24 | 7776 | 32.335 | 635.92 | −7.5067 | 0.373288 | −5.0293 | 3.1906 |
|       | 36 | 7776 | 31.894 | 635.94 | −7.5511 | 0.372737 | −4.9855 | 3.1658 |
|       | 48 | 7776 | 31.841 | 634.76 | −7.5656 | 0.372657 | −4.9751 | 3.1615 |
| *har* | 36 | 7776 | 31.022 | 683.30 | −7.0611 | 0.3722 | | 2.9615 |
|       | 48 | 13824 | 31.169 | 666.83 | −7.2790 | 0.3725 | | 3.0564 |
| *ben* | | | 30.733 | 626.41 | −7.6250 | 0.37338 | −4.9289 | 3.1017 |

case, there is a clear convergence of the results towards the benchmark estimate. In contrast to the non-magnetic case, however, the results for the drift frequency have a similar accuracy as the other requested data. This is an additional hint that the extreme sensitivity of the drift frequency in the non-magnetic benchmark case is an intrinsic feature of the specific benchmark case. For both addressed benchmark cases, the presented local approaches give reliable solutions, but they cannot compete with the accuracy of fully spectral or spectral element approaches within the rather restricted parameter regime of the benchmark study.

### 4.2.4 Summary and Future Prospects

As already pointed out, the interest in the application of local approaches to large dynamo simulations is significant. Above, we have sketched the strategies of recent implementations. A comparison with known benchmark numerical approaches, but also demonstrated that only the spectral element approach can give an accuracy comparable to fully spectral methods. This result is not surprising, since the benchmark cases are calculated at a very moderate Ekman number of $E = 10^{-3}$, i.e. at a parameter level characterized by very smooth solutions.

Much more challenging are solutions in the geophysically more interesting low Ekman number regime. This is illustrated by Fig. 4.2 that displays a snapshot of the magnetic field component $B_r$ for a solution obtained at an Ekman number of $E = 2 \times 10^{-5}$. Nearly $10^7$ grid cells have been used to calculate this particular model by the finite volume approach of Harder and Hansen (2005). The four-fold azimuthal symmetry of the benchmark cases has disappeared, and the magnetic field structures are now dominated by extremely small-scaled travelling spots. These spots are much smaller than the features found at $E = 10^{-3}$, which we will discuss further in Sect. 4.4.2. We can expect that the length scales will be even smaller when the Ekman number is decreased further. Only when the solution is filtered to spherical harmonics $\ell \leq 10$ the familiar large-scale flux bundles are recovered and the mode with the largest amplitude, the dipole component, stands out. This filtered solution is not too different from the large scale features found at $E = 10^{-3}$. It will be interesting to see whether this similarity remains at even smaller Ekman numbers.

## 4.3 Dynamos in Cartesian Geometry

Even though many characteristic features of the Earth's magnetic field have been reproduced by numerical simulations, the physics of planetary dynamos is far from being completely understood. As described above, even the most advanced simulations have to use control parameters which differ by many orders of magnitude from values that would be realistic for the Earth's core. This certainly raises the question if the simulations give a realistic picture of the dynamical processes in the Earth's

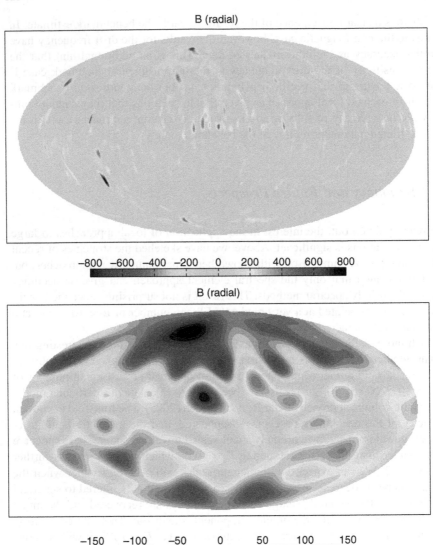

**Fig. 4.2** *Top*: Snapshot of the radial magnetic field component $B_r$ just below the upper boundary. Parameters are: $Ra = 3600$, $E = 2 \times 10^{-5}$, $Pr = 1$, $Pm = 0.45$. *Bottom*: As above, but filtered for $\ell \leq 10$

core and if the apparent agreement with geomagnetic observations (see Sect. 4.4) may perhaps only be a coincidence.

Since it will remain impossible to use realistic control parameter values in global geodynamo simulations in the near future, it seems sensible to supplement the spherical shell models with more idealized approaches. One way to do this is to abandon the spherical shell geometry and to study the dynamo processes in a Cartesian rapidly rotating Rayleigh-Bénard layer (Childress and Soward, 1972; Soward, 1974; Fautrelle and Childress, 1982; St. Pierre, 1993; Jones, 2000; Rotvig and Jones, 2002; Stellmach and Hansen, 2004; Cataneo and Hughes, 2006). Sometimes, this is motivated by considering the Cartesian domain as a representation of a small fraction of a planetary core. The restriction to a box geometry greatly simplifies the numerics and allows to study the system behavior at more realistic control parameters than can possibly be afforded in fully 3d spherical shell simulations.

The model we discuss in the following consists of a plane fluid layer that is heated from below and rapidly rotates about a vertical axis with constant angular velocity. Once more, the Boussinesq approximation is used, and we assume that gravity is aligned with the rotation axis. Accordingly, the buoyancy term is the only difference between the system equations in spherical shell geometry and in the box geometry assumed here. The Navier–Stokes equation (4.1) now reads:

$$E\left(\frac{\partial}{\partial t} + \mathbf{U} \cdot \nabla - \nabla^2\right)\mathbf{U} + 2\hat{\mathbf{z}} \times \mathbf{U} + \nabla \Pi = RaPr^{-1}T\hat{\mathbf{z}} + P_m^{-1}(\nabla \times \mathbf{B}) \times \mathbf{B}. \quad (4.15)$$

We have used the layer depth $d$ as the fundamental length scale, otherwise the scaling is identical to the one used in the spherical approach (see Sect. 4.1). The aspect ratio $\Gamma$ between the square horizontal and the vertical box dimension is an additional free parameter. Stress free, isothermal, and electrically perfectly conducting boundary conditions are assumed at the top and bottom boundaries. Periodic boundary conditions are assumed in horizontal directions.

### 4.3.1 Rapidly Rotating Rayleigh-Bénard Convection

Because of the notorious difficulties in understanding turbulent flows, the nature of the fluid motions generating the geomagnetic field is not fully clear so far. The Lorentz forces introduce a further nonlinearity into the equations, complicating the situation even more. In this section, we neglect magnetic forces entirely and use our simple Cartesian model to illustrate characteristic features of rapidly rotating convective systems. The complicated dynamical effects caused by magnetic fields will be discussed later on in a separate section.

Linear theory predicts that rotation inhibits the onset of the convective instability in a plane layer, or more precisely, that the critical Rayleigh number $Ra_c$ for the onset of convection scales as $E^{-1/3}$ in the limit $E \to 0$ (Chandrasekhar, 1961). Convection sets in as narrow cells with horizontal length scales $l_c = O(E^{1/3})$, showing non-oscillatory behavior for $Pr \geq 1$ and oscillatory behavior for small Prandtl numbers. We focus on the former case in the following. It is then possible to illustrate the linear stability results by considering the vertical component of the vorticity equation

$$E(\partial_t \omega + \mathbf{U} \cdot \nabla \omega - \omega \cdot \nabla \mathbf{U} - \nabla^2 \omega) = 2\partial_z \mathbf{U} + RaPr^{-1}\nabla \times (T\hat{\mathbf{z}}), \quad (4.16)$$

where $\omega = \nabla \times \mathbf{U}$ denotes the vorticity. Convection can only occur if either inertial or viscous contributions balance the term $\partial_z U_z$ that stems from the Coriolis force. For the Prandtl numbers considered here, viscous forces play this role. The scaling $l_c = O(E^{1/3})$ then follows directly from (4.16) as long as $\partial_z U_z = O(1)$. Small length scales of order $E^{1/3}$ are a rather universal feature of rapidly rotating convective systems. For example, such length scales are also a first order feature of convection in rotating spherical shells. Here, the critical wave number $m$ of the first unstable mode is $O(E^{-1/3})$, and the critical Rayleigh number also scales as $E^{-1/3}$, just as in the Cartesian case.

As $Ra$ is increased beyond $Ra_c$, finite amplitude convection occurs and nonlinear effects come into play. In our plane layer geometry, they cause complicated flows even in the weakly supercritical regime. In contrast to non-rotating Rayleigh-Bénard convection, where (apart from situations where defects in roll patterns are important or $Pr$ is low) stationary flows usually develop for a considerable range of supercritical Rayleigh numbers, complicated time dependent flows emerge in the rotating case. This time dependence is caused by the Küppers-Lortz instability, which leads to a heteroclinic cycle of alternating roll patterns (Busse and Clever, 1979; Busse and Heikes, 1980; Demircan et al., 2000; Jones and Roberts, 2000a; Clune and Knobloch, 1993). Since such patterns are of secondary importance for the geodynamo problem, we do not go into detail here but describe the system evolution as $Ra$ is increased further. Figure 4.3 shows visualizations of the convective flows along with time series of $Re$ and $Nu$ for $E = 2 \times 10^{-4}$, $Pr = 1$ and increasing $Ra$. For moderately supercritical Rayleigh numbers (see Fig. 4.3a), the flows are strongly influenced by rotation, and columnar structures develop whose dominant horizontal length scale $l_c$ is of the predicted order $E^{1/3}$. Vertical velocity and vorticity tend to be correlated in the lower half and to be anti-correlated in the upper half of the layer, leading to an almost antisymmetric helicity distribution with respect to the midplane.

The columnar regime persists for a finite range of Rayleigh numbers until more complex turbulent flows eventually emerge. The dominance of large vertical length scales becomes weaker with growing $Ra$, and the temporal fluctuations increase. Geostrophic motions, driven by inertial effects, interact with the convective flow and distort the columnar structures. Eventually, thermal plumes develop out of a buoyant instability of the thermal boundary layers.

The plumes form at the junction of cell-like structures in the thermal boundary layer and exhibit strong cyclonic vorticity. This cyclonicity is caused by the deflecting influence of Coriolis forces on the horizontal motions feeding the plumes (Julien et al., 1996). Weak anticyclonic vorticity persists in large regions between the thin, strongly cyclonic plume regions, such that the mean vertical vorticity vanishes when averaged over a horizontal plane.

The transition from columnar to plume dominated convection goes along with a growing ratio of buoyant to Coriolis forces. A measure for this ratio is the convective Rossby number $Ro_c = (RaE/Pr)^{1/2}$. The simulations visualized in Fig. 4.3 cover the range from $Ro_c \ll 1$ to $Ro_c = O(1)$. For $Ro_c \gg 1$, rotational effects are of secondary importance and the dynamics is essentially the same as in the nonrotating

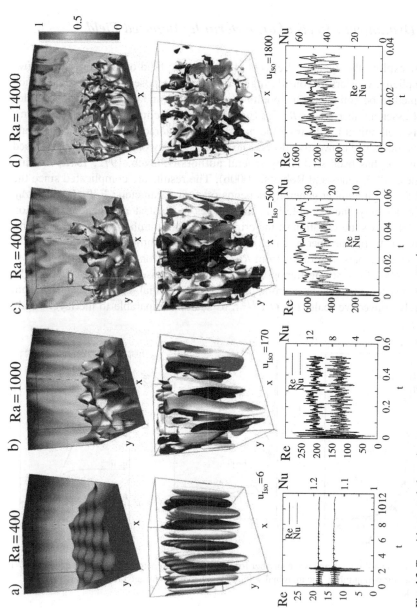

**Fig. 4.3** Transition to turbulence in non-magnetic convection at $E = 2 \times 10^{-4}$, $Pr = 1$. Shown are isosurfaces of temperature at $T = 0.7$, isosurfaces of vertical velocity at $v_z = \pm u_{Iso}$ and time series of $Re$ and $Nu$

case (Vorobieff and Ecke, 2002). Since the convective Rossby number is small in planetary cores, this regime is not relevant for the geodynamo. The Coriolis forces are strong within the Earth core, and in the absence of magnetic fields, we would assume that convection occurs in narrow cells. We will therefore focus entirely on this regime when we investigate the dynamical effect of magnetic forces on convective flows in the next section.

### 4.3.2 Dynamical Effects of an Externally Imposed Field

We now explore the dynamical effects of a magnetic field on the convective flow. For simplicity, we start by considering an externally imposed magnetic field which is assumed to be homogeneous, points in $x$-direction, and whose amplitude scales with Elsasser number $(\Lambda_0 = B_0^2/2)$. Even though this configuration is rather simple, it is well suited to illustrate the nature of the interaction between large scale magnetic fields and convective flows. The linear stability problem for the described configuration has been studied by several authors (Eltayeb, 1972, 1975; Roberts and Jones, 2000; Jones and Roberts, 2000b). The results are complicated since the critical Rayleigh number generally depends on four parameters: $E, Pr, Pm$, and $\Lambda_0$. We consider only one example here that serves to illustrate the general behavior. Figure 4.4 shows the variation of the critical Rayleigh number $Ra_c$ and wave number $k_c$ with $\Lambda_0$ for the special case $q = \kappa/\lambda = 1$ and infinite Prandtl number. At low Ekman numbers, $Ra_c$ drops from $Ra_c = O(E^{-1/3})$ for weak magnetic fields to $Ra_c = O(100)$ for a strong imposed field with an Elsasser number of $O(1)$ . The transition is accompanied by a strong increase of the spatial scales which grow from the purely convective scale, $l = O(E^{1/3})$, to scales comparable to the layer depth,

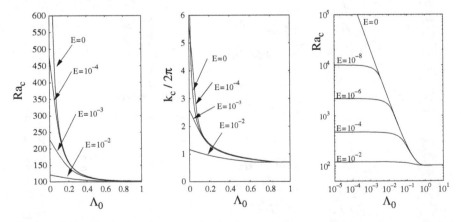

**Fig. 4.4** Critical Rayleigh and wave number for the onset of magnetoconvection for $q := \kappa/\eta = 1$ and infinite Prandtl number. The marginal velocity field components are proportional to $\exp(i\,\mathbf{k}\cdot\mathbf{r})$ and $k = |\mathbf{k}|$

$l = O(1)$. Magnetic forces help to offset the rotational constraint provided $\Lambda_0$ is large enough. They contribute in balancing the term $\partial_z U_z$ in Eq. (4.16), which had to be balanced solely by viscous forces in the non-magnetic problem. For strong enough imposed fields, viscous forces become irrelevant and the need for small length scales disappears completely.

The linear stability results discussed above illustrate that a pronounced separation of weak and strong field states is to be expected only for $E \ll 1$. In the following, we discuss numerical simulations at $E = 10^{-5}$ where the pure convective and the magnetically controlled regime can clearly be distinguished. The remaining parameters, $Pr = 1, Ra = 1200$, and $q = 0.5$, have been chosen to exclude kinematic dynamo action and also to enable a direct comparison with the non-magnetic columnar solution discussed in the previous section. Figure 4.5 shows two solutions for a weak and a strong imposed field at $\Lambda_0 = 10^{-3}$ and $\Lambda_0 = 10$, respectively. At $\Lambda_0 = 10^{-3}$, the Lorentz force has negligible influence on the flow dynamics and the solution corresponds to the purely convective columnar regime. At $\Lambda_0 = 10$, the solution is drastically different. The convective columns have vanished and a generally large scale convection prevails. Thin thermal boundary layers develop which indicates that the large scale flow is very efficiently transporting heat. Accordingly, the time averaged Nusselt number is about an order of magnitude larger than in the weak field case.

Further insight into the dynamics is gained by comparing the Ohmic and viscous dissipation rates: Viscous effects dissipate nearly all the energy fed into the system when the imposed field is small, whereas Ohmic dissipation dominates in the large $\Lambda_0$ case. Nearly all kinetic energy released by the buoyancy force is then transferred

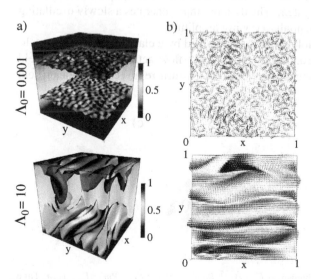

**Fig. 4.5** Snapshots of magneto-convective flows at $E = 10^{-5}$, $Pr = \Gamma = 1$, $Pm = 0.5$ and $Ra = 1200$ for weak and strong imposed horizontal fields. (a) Isosurfaces of temperature for $T = 0.3$ and $T = 0.7$. (b) Velocity field at the upper boundary illustrated by vectors that have been scaled with the local flow velocity

to the magnetic energy through the stretching of magnetic lines of force by the fluid motions. The electric currents generated in this way experience Ohmic losses and finally transfer the energy into heat. Viscous dissipation is significantly reduced by the increase in flow scale and becomes almost negligible in comparison to the Ohmic losses.

Simulations at larger values of $E$ confirm that the differences between strong and weak field solutions gradually vanish with increasing Ekman number, and we can speculate whether dynamo models running at too large Ekman numbers possibly miss this important dynamical effect.

### 4.3.3 Self-consistent Dynamos

We now move on to the full dynamo problem where the magnetic field is not imposed externally but generated by the fluid flow itself. Small magnetic disturbances start to grow exponentially once the Rayleigh number has been increased to a critical value that lies beyond the value for the onset of convection. Figure 4.6 illustrates the field morphology during the initial growth phase. The magnetic field is characterized by two preferred spatial scales: a small scale component that varies on the same horizontal scales as the columnar convective flow and a strong mean field component, which is defined as the horizontally averaged magnetic field in this context. Both field scales are clearly evident in the magnetic energy spectrum shown in Fig. 4.6b. The dominating mean field has a spiral staircase structure that slowly rotates in a sense opposite to the system rotation on a time scale comparable to the free magnetic decay time of the system. The dynamo thus generates a slowly oscillating dynamo wave with an exponentially growing amplitude.

The observed field morphology can be explained by a classical two-scale mechanism (Childress and Soward, 1972): Small scale flow acting on the mean field generates the small scale magnetic field which in turn reinforces the mean field

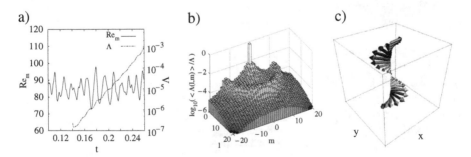

**Fig. 4.6** Kinematic dynamo mechanism at $E = 10^{-5}, Ra = 1200$ and $Pr = Pm = \Gamma = 1$. (**a**) Time series of $Re_m$ and $\Lambda$. (**b**) Spectrum of magnetic energy, i.e. the magnetic energy $\Lambda(l,m)$ contained in modes with wave number $\mathbf{k} = 2\pi(l\mathbf{e_x} + m\mathbf{e_y})/\Gamma$ normalized by the entire magnetic energy $\Lambda$ and averaged over the time span of kinematic growth. (**c**) Structure of the horizontally averaged magnetic field

through an interaction with the small scale flow. This cycle manages to overcome the Ohmic decay and leads to the observed exponential field growth. The process of creating a mean field by the interaction of the two small scale constituents is called an $\alpha$-effect in the framework of classical mean field magnetohydrodynamics (Krause and Rädler, 1980).

The exponential growth stops once the magnetic field is strong enough to modify the convective flow via the Lorentz force. Different scenarios have been debated how this magnetic field saturation may be achieved. We follow the so-called strong-field scenario here in applying the results from the magneto-convection studies presented in the previous section to the self-consistent case. Both, the linear studies and the fully nonlinear simulations of magneto-convection discussed above, suggest that once the magnetic field has reached a certain intensity, the convective flow may gain in strength, which in turn causes an increasing growth rate of the magnetic energy. For $E \ll 1$, a runaway effect is to be expected, in which the magnetic field grows faster than exponentially until the Elsasser number becomes $O(1)$, and the system saturates. Analytical investigations (Soward, 1974; Fautrelle and Childress, 1982) based on an amplitude expansion at small $E$ support these somewhat heuristic ideas.

A clear separation of weak and strong field states can be expected only for $E \ll 1$, a parameter regime difficult to investigate numerically. Simulations by Stellmach and Hansen (2004) reveal that the promoting effect of the magnetic field on the convective flow can already be observed at $E = 10^{-5}$. Figure 4.7 illustrates

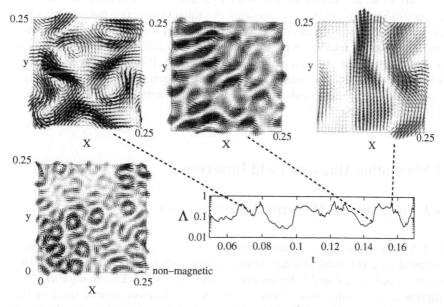

**Fig. 4.7** Velocity field at the upper boundary $z = 1$ for $E = 10^{-6}$, $Ra = 2400$, $Pr = Pm = 1$ visualized by arrows that have been scaled with the local flow speed. The graph in the *lower right* shows the time history of the Elsasser number and *the dashed lines* indicate the time instants at which the different snapshots have been taken. For comparison, the non-magnetic case is shown in the *lower left panel* (after Stellmach and Hansen, 2004)

the flow structure in the magnetically saturated regime for $E = 10^{-6}$, $Ra = 2400$, $Pr = Pm = 1$, and $\Gamma = 0.25$. It differs strongly from the non-magnetic case. On time average, the flow has higher kinetic energy, is dominated by larger spatial structures, and transports heat more efficiently. An inspection of the temporal behavior reveals that strong large scale flows develop during episodes of intense magnetic field, while more organized small scale convection arises during times of lower field intensity. Somewhat surprisingly, all dynamos found by Stellmach and Hansen (2004) saturate at $\Lambda < 1$. This field strength is too small to cause a quasi-magnetostrophic state similar to the one observed in the magneto-convection simulations at $\Lambda_0 = 10$ (see Sect. 4.3.2).

The Cartesian simulations discussed so far have employed moderate Rayleigh numbers, and we may speculate whether the magnetostrophic state may be reached at higher Rayleigh numbers where a stronger magnetic field can be expected. This idea is supported by numerical simulations at infinite Prandtl number by Rotvig and Jones (2002), who found nearly magnetostrophic dynamos for the special case of rigid and electrically insulating boundary conditions. Our first simulations at $Pr = 1$, however, reveal that the more complex flow at higher $Ra$ typically leads to a breakdown of the simple scale separation instrumental in the two-scale mechanism. There is an intermediate regime where dynamo action breaks down altogether. At larger Rayleigh numbers, the flow once again picks up dynamo action but generates a small scale fast fluctuating field without a significant large scale component. The situation thus differs strongly from the magneto-convective situation discussed in Sect. 4.3.2. The failure to act as an efficient large scale dynamo is somewhat unexpected since the flow still has a strongly helical character. A similar observation has recently been made by Cataneo and Hughes (2006) who studied high Rayleigh number dynamos at moderate Ekman number. Why the $\alpha$-effect is so inefficient despite the helical nature of the convective flows remains an open question. It is interesting to compare this behavior to the high $Ra$ cases in spherical shell dynamos (see Sect. 4.4.2), where the production of the large scale magnetic field also breaks down. We will discuss this issue further in Sect. 4.5.

## 4.4 Simulating Magnetic Field Reversals

### 4.4.1 Fundamental Magnetic Field Structure

We start with outlining the dynamo action that leads to the fundamental dipole-dominated magnetic field structure in our spherical shell models. The complexity and time dependence found in higher supercritical and/or low Ekman number cases complicates the identification of key processes. We therefore concentrated on the benchmark II that is characterized by a particularly large scale solution with a fourfold azimuthal symmetry (Christensen et al., 2001). It also offers an intriguingly simple time behavior that we refer to as quasi-stationary: the only time dependence

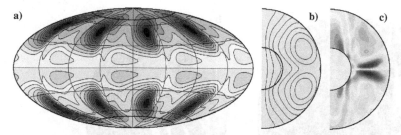

**Fig. 4.8** Magnetic field for the benchmark II dynamo (Christensen et al., 2001). Panel a) shows the radial magnetic field at the outer boundary, blue marks ingoing field. Panel b) shows the azimuthally averaged field lines, and panel c) displays the axisymmetric radial magnetic field production, i.e. the radial component of the induction term $\nabla \times (\mathbf{U} \times \mathbf{B})$

is a drift of the whole solution in azimuth. The parameters are $E = 10^{-3}$, $Ra = 110$, $Pr = 1$, and $Pm = 10$. Analysis at more extreme parameters, in particular smaller Ekman numbers and larger Rayleigh number, suggest that many of the features identified in the simple benchmark dynamo still apply (Olson et al., 1999).

Figure 4.8 shows the radial magnetic field at the outer boundary, the axisymmetric field lines, and the radial component of the induction term $\nabla \times (\mathbf{U} \times \mathbf{B})$, which is perfectly balanced by Ohmic decay in a quasi-stationary case. The underlying dynamo process, responsible for shaping the magnetic field, has been envisioned by Olson et al. (1999) and was later confirmed in a detailed analysis by Wicht and Aubert (2005). Figure 4.9 shows iso-surfaces of the axial vorticity ($z$-component) that highlight the columnar structure of the convective flow. Anti-cyclones (shown in blue), that rotate in the direction opposite to the system rotation, are significantly stronger than cyclones (red) and play a key role in the dynamo process. Figure 4.9 visualizes the anti-cyclonic action with the help of magnetic field lines. The (imaginary) field production cycle starts in the equatorial region where the radial outflow grabs a north–south oriented field line that represents the dominant dipole direction. Radial stretching of the field line produces strong inverse radial field on either side of the equatorial plane (see Fig. 4.8c). The pairwise equatorial field patches often found at the outer boundary of many dynamo simulations are reflections of this internal process.

The next steps in the production cycle are the wrapping of field lines around the anticyclone and its advection in northward and southward direction on both sides of the equator, respectively. Responsible for the latter action are meridional flows, which are roughly directed away from the equatorial plane in anticyclones but converge at the equatorial plane in cyclones. These relatively weak flows are vital for the dynamo process, since they separate the opposing radial field that largely cancels in the equatorial region. The diverging axial flows in anticyclones also produce magnetic field by stretching the field line in the north–south direction. The cycle ends with another north–south oriented field line that amplifies the starting line and thereby compensates Ohmic decay.

Flows converging where cyclones come close to the outer boundary further shape the magnetic field by advectively concentrating the background dipole field. These

a)    b)

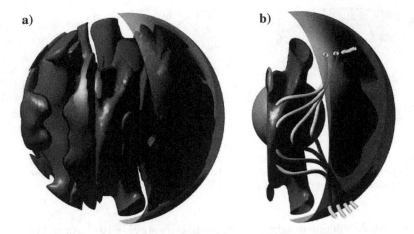

**Fig. 4.9** Flow structure and dynamo process in the benchmark II dynamo. Panel (**a**) shows positive and negative iso-surfaces of axial vorticity in red and blue, respectively. Anti-cyclones, the structures of negative axial vorticity, dominate magnetic field production. Their role in the dynamo process is visualized in panel (**b**) with magnetic field lines shown in yellow. The field production cycle starts with a roughly north–south oriented field line at right that is stretched outward in the equatorial region, and then proceeds towards the left

flows are part of the meridional flow that continues along the cyclonic axis towards the equator. The inverse field production inside the tangent cylinder, shown in (see Fig. 4.8c), largely reflects this advective weakening of the normal polarity field by the meridional circulation that is directed away from the rotation axis.

Pairs of inverse fields around the equator, stronger normal polarity patches at higher latitudes, and a weaker radial field around the poles, all theses features are also prominent in the geomagnetic field (see Fig. 4.13), which suggest that at least some aspects of the outlined fundamental dynamo mechanism may also apply to the geodynamo (Olson et al., 1999). However, we have described a quasi-stationary dynamo that neither shows the secular variation observed on Earth, nor will it ever reverse. Time variability and polarity reversals appear when we increase the Rayleigh number in our system. In the following section, we try to pin down the parameter space where reversals can be found and set out to determine the necessary flow properties for reversals to happen.

## 4.4.2 Stable and Reversing Dynamos

The eminent success of the work by Glatzmaier and Roberts (1995) is partly owed to the fact that they presented the first numerical dynamo with an Earth-like reversal behavior. The attribute Earth-like refers to a first order property of geomagnetic polarity changes: they are rare. Long stable polarity epochs, lasting between some ten thousand to several million years, are separated by significantly shorter transitional reversal periods, lasting from a few thousand to ten thousand years.

The reversing dynamo models preceding the work by Glatzmaier and Roberts (1995) were kinematic dynamos that continuously switch polarity and show no intervening stable polarity epochs. Rather than solving the Navier–Stokes equation, kinematic dynamo models prescribe the flow field so that the induction equation Eq. (4.2) becomes an eigen-value problem. Exponentially growing (or marginally stable) oscillatory eigen-solutions correspond to reversing dynamo cases. Whether or not a solution is oscillatory is determined by the prescribed flow alone, and several authors explored a multitude of simple parameterized flow fields in order to pin down the flow features that promote or suppress field reversals (See Gubbins et al., 2000a,b, and references therein). Meridional circulation, for example, was found to have a stabilizing effect (Gubbins and Sarson, 1994; Sarson and Jones, 1999). The conclusions drawn from kinematic dynamo models, however, may not translate to fully self-consistent dynamos. Wicht and Olson (2004), for example, observe that meridional circulation is essential for the reversal mechanism in their model, despite the fact that their simulation comes close to a kinematic case.

Willis and Gubbins (2004) modify a kinematic model in order to explain the changeover from shorter transitional to longer stable epochs by prescribing a gradual change in the flow pattern. While this approach falls short of explaining the dynamical processes that lead to reversals, it nevertheless points in the right direction: For a stationary flow, the magnetic polarity is either stable or oscillates continuously. Flow changes are thus the only way to explain the existence of longer stable polarity epochs and the irregular appearance of reversals.

Kutzner and Christensen (2002) studied the transition from stable to reversing dynamos by increasing the Rayleigh number for several parameter combinations. They identified three different regimes: At lower Rayleigh numbers, the dipole never reverses and by far dominates the magnetic field. At large Rayleigh numbers, the dipole reverses continuously but never clearly dominates. We denote these regimes with SR for stable regime and MR for multipole regime (or multiple reversals), respectively. Regime ER, where Earth-like excursions and reversals occur, lies where SR and MR meet, Earth-like in the sense that transitional periods are rare and that the dipole still dominates in a time-mean sense.

SR and ER form the regime of dipole dominated dynamos DR. Christensen and Aubert (2006) as well as Olson and Christensen (2006) analyzed a larger suit of dynamos than had been explored by Kutzner and Christensen (2002) and conclude that the boundary between regimes DR and MR is best characterized in terms of the local Rossby number $Ro_\ell = U/\ell\Omega$, where $\ell$ is a measure for the latitudinal flow scale that we introduced in Sect. 4.1. They find that the transition from dipolar to multipolar dynamos takes place at $Ro_\ell \approx 0.12$ and suggest that regime ER can be found towards lower $Ro_\ell$ values but not too far away from the boundary. This inference nicely goes along with the fact that estimates of Earth's local Rossby number amount to $Ro_\ell \approx 0.09$ (Olson and Christensen, 2006).

Since they focussed on exploring a larger range of parameters and on mapping out the boundary between dipolar and multipolar dynamos, Kutzner and Christensen

(2002), Christensen and Aubert (2006) and Olson and Christensen (2006) could not afford the longer runs necessary for analyzing the reversal behavior and to explore the actual extension of regime ER. We complement their studies in this respect by concentrating on fewer cases at not too small Ekman numbers and by more closely mapping out the regimes for three dynamo families. These families differ in Ekman number, magnetic Prandtl number, and/or driving mechanism, and we gradually increase the Rayleigh number to track the transition from SR to MR. Representatives of each family are listed in Table 4.3. Family T and C run at an Ekman number of $E = 10^{-3}$ but have different driving mechanisms. Model family T is driven by an imposed buoyancy contrast, while family C employs what Kutzner and Christensen (2002) call a chemical convection model: The buoyancy flux is forced to vanish at the outer boundary, and a constant buoyancy is imposed at the inner boundary. The same driving type is also used for the lower Ekman number cases at $E = 3 \times 10^{-4}$, that further explore a parameter combination introduced by Kutzner and Christensen (2002); Table 4.3 lists models C1 and KC02 to represent the two chemical convection models.

Two measures serve to distinguish whether a dynamo model shows sufficiently Earth-like reversal behavior: the relative transitional time $\tau_T$ and the degree of dipole dominance $D$. We define magnetic pole positions further away than $\delta \Theta_c = 45°$ from the closest geographical pole as transitional here. Note that 'magnetic pole' refers to the geomagnetic north pole established by the dipolar field contributions alone. $\tau_T$ is the fraction of time the magnetic pole spends at transitional positions and helps to access whether stable periods are still the norm. The degree of dipole dominance

**Table 4.3** List of parameters and properties for the dynamos analyzed. Magnetic Reynolds number $Rm$, Elsasser number $\Lambda$, local Rossby number $Ro_\ell$, true dipole moment TDM, and relative CMB dipole strength $D$ have been averaged over longer representative sequences. The Prandtl number is unity in all cases. The last column list the relative transitional dipole time $\tau_T$, i.e. the fraction of time the magnetic north pole spends further away from the closest geographic pole than 45°. Column 4 lists the driving mechanism: an imposed constant temperature jump across the shell (temp.), or chemical convection (chem.) in the sense that the boundary conditions allow no buoyancy flux through the outer boundary. For Earth we list an Ekman number and a magnetic Prandtl number that are based on molecular diffusivities. Earth's core Rayleigh number is hard to estimate but thought to be highly supercritical (Gubbins et al., 2004). See main text for more explanations

| Name | $E$ | $Pm$ | BC | $Ra$ | $Ra/Ra_c$ | $\overline{Rm}$ | $\overline{\Lambda}$ | $Ro_\ell$ | $\overline{TDM}$ | $\overline{D}$ | $\tau_T$ |
|------|-----|------|------|------|-----------|------|------|------|------|------|------|
| T1 | $10^{-3}$ | 10 | temp. | 100 | 1.8 | 69 | 11 | 0.01 | 53 | 0.7 | 0.00 |
| T2 | | | | 375 | 6.7 | 328 | 12 | 0.08 | 18 | 0.4 | 0.00 |
| T3 | | | | 450 | 8.1 | 396 | 10 | 0.11 | 12 | 0.3 | 0.02 |
| T4 | | | | 500 | 8.9 | 435 | 9 | 0.12 | 9 | 0.2 | 0.04 |
| T5 | | | | 750 | 13.4 | 591 | 13 | 0.17 | 5 | 0.1 | 0.34 |
| C1 | $10^{-3}$ | 10 | chem. | 1250 | 23.5 | 440 | 7 | 0.11 | 7 | 0.2 | 0.05 |
| KC02 | $3 \times 10^{-4}$ | 3 | chem. | 9000 | 160 | 495 | 3 | 0.16 | 3 | 0.2 | 0.15 |
| W05 | $2 \times 10^{-2}$ | 10 | temp. | 300 | 2.5 | 92 | 3 | 0.22 | 7.3 | 0.4 | 0.06 |
| Earth | $10^{-14}$ | $10^{-5}$ | | | $\gg 1$ | 500 | $O(1)$ | 0.09 | 8 | $< 0.6$ | $< 0.10$ |

$D$ is measured by the square root of the ratio of magnetic dipole energy to the total magnetic energy at the core–mantle boundary (CMB).

Estimating typical values of $\tau_T$ and $D$ for Earth is difficult, but we are only interested in a rough categorization here. $\tau_T$ probably amounts to no more than a few percent (but note Lund et al., 1989), we regard dynamos with $\tau_T \leq 0.1$ as showing Earth-like separation of stable and transitional epochs. Using recent geomagnetic field models (Maus et al., 2006), which provide spherical harmonic contributions up to degree and order 14, we estimate a geomagnetic dipole dominance of $D = 0.6$. Dynamo simulations suggest that the modes with $l > 14$ carry at least as much magnetic energy as the modes $l \leq 14$ (Wicht et al., 2007), which translates to a value of $D = 0.4$ or smaller. Paleomagnetic data and dynamo simulations agree in the fact that $D$ is decisively lower during transitional periods. Time averaged values $\overline{D}$ are thus lower than values based on stable polarity epochs alone, an effect that grows with $\tau_T$. We regard dynamo models with $\overline{D} \geq 0.2$ to work in the Earth-like regime ER here.

Table 4.3 gives values of $\tau_T$ and $\overline{D}$ for some of the dynamo models studied and also lists the time averaged true dipole moment (TDM). The adjective 'true' serves to distinguish the true spherical harmonic dipole contribution from the virtual dipole moment commonly used in local paleomagnetic analysis (see below). An estimate of Earth's TDM based on recent field models for the year 2000 (Maus et al., 2006) provides a value of $8 \times 10^{22}\,\mathrm{Am}^2$. However, the TDM has likely been lower in the past (Olson and Amit, 2006). We note that Elsasser numbers as well as magnetic Reynolds numbers are Earth-like for all three dynamo families explored in more detail here. Elsasser numbers of order one establish what is called the magnetostrophic regime thought to rule the dynamics in planetary iron cores. The magnetic Reynolds number measures the ratio of magnetic field production to magnetic field diffusion and is therefore a second essential parameter for the dynamo process. The fact that Elsasser number and magnetic Reynolds number are Earth-like is probably the key reason for the success of dynamo models in explaining geomagnetic features (Christensen and Wicht, 2007).

Figure 4.10 shows regime diagrams of the Kutzner and Christensen (2002) type (panel a) and of the Christensen and Aubert (2006) type (panel b). In panel a) we have supplemented a figure by Christensen and Aubert (2006) with cases of dynamo family T. Panel b) covers all three families and in addition model W05. Note that Kutzner and Christensen (2002) and Christensen and Aubert (2006) drew the boundary between DR and MR as an upward continuation of the cusp in the convective regime CR in panel a). Our analysis, however, suggests that the transition takes place at higher Rayleigh numbers. We have chosen a magnetic Prandtl number of $Pm = 10$ in order to avoid the convective cusp.

Figure 4.10b demonstrates that $\overline{D}$ decreases rather smoothly with increasing $Ro_l$ in all cases. This is particularly evident for the two model families T and C at $E = 10^{-3}$ that show nearly identical behavior (squares and circles in 4.10b). The similarity of both curves also confirms that the regime categorization in terms of $\overline{D}$ and $Ro_\ell$ is insensitive to the two convective driving types explored here. Note, however, that dynamos driven by internal heat sources that are homogeneously

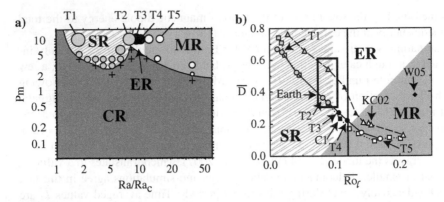

**Fig. 4.10** Regime diagrams showing the stable dipolar regime (SR, *scribbeld*), Earth-like reversing regime (ER, *white*), the multipolar regime (MR, medium grey), and the purely convective regime (CR, *dark grey*). Panel (**a**) shows a regime diagram following Kutzner and Christensen (2002) and extends a plot by Christensen and Aubert (2006) with the dynamo family T. The size of the symbols is a rough measure for the dipole dominance $\overline{D}$. Panel (**b**) follows the regime diagrams by Christensen and Aubert (2006) and Olson and Christensen (2006). *Circles* and *rectangles* indicate dynamos of family $T$ and $C$, respectively. *Triangles* mark the dynamos at $E = 3 \times 10^{-4}$ which are represented by case KC02 in Table 4.3; *the single diamond* marks model W05. *Grey symbols* stand for non-reversing cases, black symbols mark Earth-like reversing dynamos with $\tau_T \leq 0.1$, and *open symbols* indicate more frequently reversing cases with $\tau_T > 0.1$. The *black rectangle* shows the parameter range of estimated geodynamo values

distributed over the shell tend to produce less dipole dominated fields and not strictly follow the outlined $Ro_\ell$ regime diagram (Kutzner and Christensen, 2000; Olson and Christensen, 2006).

The value $Ro_\ell \approx 0.12$, which has been suggested as the boundary between regimes DR and MR by Christensen and Aubert (2006), indeed marks where the dynamo families change from the dipolar to the multipolar regime. The boundary also roughly coincides with the transition from Earth-like rare to more frequent transitional events at $\tau_T = 0.1$. This was actually one of the motivations to chose this particular $D$ and $\tau_T$ parameter combination that defines regime ER. In fact, all dynamos with $\tau_T \leq 0.1$ also fulfill our second requirement $D \geq 0.2$. Regime ER is rather narrow with a width that corresponds to an increase of the Rayleigh number by roughly 50% for all three dynamo families. The regime widening apparent in Figure (4.10b) primarily reflects our lack of knowledge, we should keep in mind that the regime width merely depends on our somewhat arbitrary categorization. For example, if we accept values up to $\tau_T = 0.15$, regime ER extends from $Ro_\ell \approx 0.1$ to $Ro_\ell \approx 0.15$ for the dynamos at $E = 3 \times 10^{-4}$, which corresponds to an increase in Rayleigh number by a factor of two. This widening of regime *ER* largely relies on the fact that the dependence of $\overline{D}$ on $Ro_\ell$ becomes weaker for larger $\overline{D}$ values. A steeper decrease in $\overline{D}$ beyond $Ro_\ell = 0.12$ and the subsequent levelling-off may reflect the sharper regime separation observed by Christensen and Aubert (2006) and Olson and Christensen (2006). It also seems to roughly support our choice of critical values for $\tau_T$ and $\overline{D}$ and is in favor of a narrow ER regime. The black rectangle in Figure (4.10b) represents Earth parameters, and its width gives an idea of potential uncertainties. The value $Ro_\ell = 0.9$, derived by Olson and Christensen (2006), is based on a scaling that involves the Earth's CMB heat flux as the least well known quantity. We have assumed that this heat flux is known up to a factor two, but the uncertainty may be larger. Reversing dynamo models are found for $Ro_\ell \geq 0.1$, and cases on the low $Ro_\ell$ side may coincide with Earth parameters in terms of $Ro_\ell$, $D$, and TDM.

We can only speculate why $Ro_\ell$ has to exceed a critical value for reversals to happen. $Ro_\ell$ is a measure for the ratio of inertia to Coriolis force in the Navier–Stokes equation (4.1). A dominant Coriolis force has a strongly ordering influence on the flow pattern and is responsible for the so-called geostrophic configuration (see Sect. 4.3). This configuration, where the flow varies only little along the $z$-axis, is typical for strongly rotating low-viscosity systems where the Coriolis force is largely balanced by the pressure gradient. The flow coherence imposed by the Coriolis force is ultimately responsible for the dipole dominance at low Rayleigh numbers. If other forces contribute to balance the Coriolis force, more unordered flow and magnetic field pattern can evolve and potentially lead to reversals. Why, however, reversals should happen when inertia balances roughly 10% of the Coriolis force remains unclear.

We illustrate solutions typical for regimes SR, ER, and MR with dynamo models $T1$ to $T5$. Figure 4.11 demonstrates how magnetic field and flow field geometry develop when the Rayleigh number is increased. Figure 4.12 shows the corresponding time dependence of the dipole tilt $P$, the TDM, and the relative dipole strength $D$. At larger magnetic Prandtl numbers and not too small Ekman numbers, dynamo action sets in for Rayleigh numbers where the flow still obeys a simple azimuthal symmetry and is quasi stationary. The benchmark dynamo, explored in Sect. 4.4.1, is an example for this type of solutions. A rise in $Ra$ leads to symmetry breaking in time and space: azimuthal symmetry as well as equatorial symmetry are lost, and the time dependence becomes chaotic, possibly via an oscillatory regime (Wicht, 2002). Figure 4.11 T1 depicts the solution for case $T1$ at $Ra = 4.47 Ra_c$ where a simple threefold azimuthal symmetry still clearly dominates. Flow and magnetic field show a nearly perfect equatorial mirror symmetry and antisymmetry, respectively. This model is close to the benchmark dynamo in character and is safely located in regime SR.

When increasing the Rayleigh number to $Ra = 6.71 Ra_c$ in model T2, azimuthal and equatorial symmetries are significantly broken, and the time dependence is clearly irregular. Convective motions now also fill the tangent cylinder, which was largely void of motion at the lower Rayleigh number case. Solution T2 is still clearly dominated by the axial dipole contribution, and there are no reversals nor excursions in the computational run. This dynamo is still a member of regime SR but lies close to the transition to regime ER.

Figure 4.11 T3 and Figure 4.12 T3 show the solution at $Ra = 8.1 Ra_c$. Stable polarity epochs with dominating axial dipole contributions are now separated by much shorter polarity transitions and excursions. The dipole is weak during these events but still clearly dominates most of the time with $\overline{D} = 0.3$. The relative transitional

$B_r(r = r_0)$                                    $V_r(r = r_i + (r_0 - r_i)/2)$

T1)  Ra = 4.5 Ra$_c$

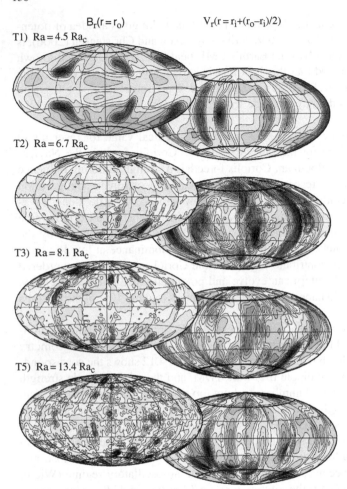

T2)  Ra = 6.7 Ra$_c$

T3)  Ra = 8.1 Ra$_c$

T5)  Ra = 13.4 Ra$_c$

**Fig. 4.11** Snapshots of the radial magnetic field component $B_r$ at the CMB ($r = r_0$) and radial flow $V_r$ at mid depth in the spherical shell for models with four different Rayleigh numbers; $r_i$ denotes the inner core radius. The cases correspond to the dynamos models listed in Table 4.3, dynamos T3 and T4 are reversing (see Figure 4.12). Blue (*yellow/red*) color indicates inward (*outward*) directed flow and magnetic field

time is $\tau_T = 0.03$ so that both values place this dynamo in regime ER. The TDM seems somewhat high at TDM=12.7 Am$^2$ compared to typical Earth values. Model *T5*, depicted in Fig. 4.11 T5 and Fig. 4.12 T5, illustrates a solution in regime MR at $Ra = 13.4 Ra_c$ . The dipole is now significantly weaker ($\overline{D} = 0.1$) and reverses persistently ($\tau_T = 0.34$).

We have already mentioned the broken equatorial and azimuthal symmetries, but there are more important differences in the magnetic field structure of solutions T1 and T2 (see Fig. 4.11). Normal as well as inverse polarity flux patches are considerable smaller in the higher *Ra* case. The mid-latitudinal normal patches now

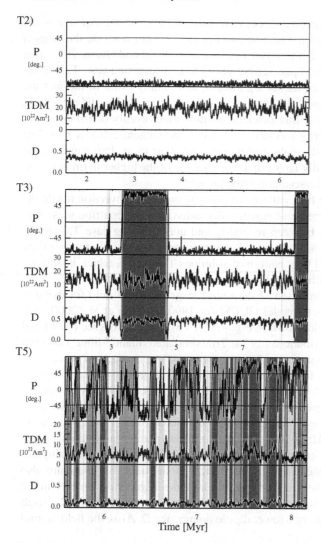

**Fig. 4.12** Dipole tilt P, true dipole moment TDM, and relative RMS dipole strength D at the CMB at three different Rayleigh numbers. The respective parameters are listed in Table 4.3. Normal and reversed polarity epochs are marked by *white and dark grey* background. A *light grey* background highlights excursions and reversals when the magnetic pole resides between 45° north and south or does not seem to settle at a stable polarity. Snapshots of corresponding magnetic and flow fields are shown in Fig. 4.11

cluster at somewhat higher latitudes, gathering just outside the tangent cylinder. As in the benchmark case, they are produced by the flows converging into downstreams along the axis of cyclones, that are now more closely attached to the inner core than in the benchmark case or for model T1. These patches establish most of the dipole field, therefore constitute the dipole dominance, and also determine the field polarity.

The equatorial pairs of inverse patches are located closer to the equator at the higher *Ra* models, and additional inverse patches start to appear inside the tangent cylinder, being produced by hot (buoyant) up-streams (Wicht and Olson, 2004) that now also fill this region. Discrepancies between the non-reversing case T2 and reversing case T3 are marginal, hardly significant, and mainly owed to the fact the we compare snapshots. Model T5, however, shows a remarkably different field structure. Normal and inverse flux patches are now more evenly distributed over the globe. The normal polarity patches just outside the tangent cylinder have lost their prominence, their north–south correlation, and their clear dominance in the azimuthal mean. The generally larger time dependence of the higher harmonic contributions leads to a strongly varying multipolar field.

The main changes in the flow pattern concern the onset of convection inside the tangent cylinder and an increase of equatorially antisymmetric contributions. And once again, the differences between reversing and non-reversing case T2 and T3 are only marginal (see Fig. 4.11). Somewhat surprisingly, we also find only minor changes between the overall flow pattern of solutions T3 and T5, which is difficult to consolidate with the significantly different magnetic field structures.

Figure 4.13 shows a comparison of the radial magnetic CMB field for dynamo T3 with the geomagnetic field model GUFM for 1990 (Jackson et al., 2000). We have filtered the dynamo field shown in Fig. 4.11 T3 by damping degree $l = 13$ and $l = 14$ and suppressing contributions $l > 14$. This simulated 'mantle-filter' eases the comparison with the geomagnetic field where the crustal magnetization prevents us from knowing the magnetic field beyond approximately degree $l = 14$. A comparison of damped (see Figure 4.13) and full solution (see Fig. 4.11 T3) reveals that the lower resolution averages over underlying smaller field patches and amplifies their size, but it still captures the essential features. The differences are more pronounced at lower Ekman numbers. High latitude normal polarity patches as well as inverse patches around the equator and inside the tangent cylinder are located at comparable location in the geomagnetic and the numerical model field. But there are also differences: The magnetic field at lower to mid latitudes is generally weaker in the simulation than in the geomagnetic field. Its strength increases at higher Rayleigh numbers but only at the costs of a lower dipole dominance $D$. Also, the field around

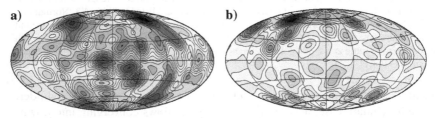

**Fig. 4.13** Comparison of the radial CMB field in the GUFM geomagnetic field model for 1990 (Jackson et al., 2000) in panel (**a**) with a snapshot of dynamo T3 in panel (**b**). The field of T3 has been damped to simulate the coarser resolution of geomagnetic models, the full numerical resolution is displayed in Fig. 4.11 T3

the equator is of predominantly normal polarity in the geomagnetic field but mainly inverse in typical reversing dynamo simulations.

The sequence for dynamo T3 shown in Fig. 4.12 demonstrates that a low dipole moment as well as a low relative dipole strength are typical for excursions and reversals (Kutzner and Christensen, 2002). Time averaged spectra, shown in Fig. 4.14, confirm that the dipole component has lost its dominance in the multipolar regime represented by case T5. Its energy content is even lower than that of the higher harmonics. The dynamo has apparently lost its capability to generate a large scale coherent dipole field. Time averaged spectra of transitional periods in model T3, excursions as well as reversals, show the same characteristics, which suggests that both types of event are essentially short interludes into a multipolar state that the dynamo assumes on time average at high Rayleigh numbers (Kutzner and Christensen, 2002). In other words: the likelihood for and the duration of these interludes increases with Rayleigh number, which is reflected by the increase of relative transitional time $\tau_T$ documented in Table 4.3.

The numerical run for the reversing dynamo shown in figure Fig. 4.12 T3 hints on the difficulties to establish whether a dynamo is reversing or remains stable. Simulated reversals like geomagnetic reversals are stochastic events with considerable variation in the duration of stable polarity epochs. The third polarity interval shown in Fig. 4.12 T3 lasts about 2 Myr, which corresponds to roughly 7 million time steps in the numerical simulation. Had we started the simulation at the beginning of this polarity epoch and had we simulated 'only' 6 million time steps, we would have concluded that this is a non-reversing case. Model $T2$ has therefore been computed for a significantly longer time span, covering nearly 5 Myr years with more than 17 million time steps. This could only be afforded for a model where the required

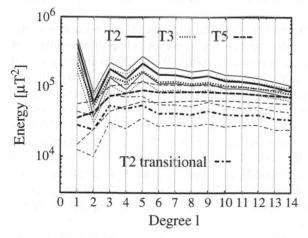

**Fig. 4.14** Time averaged magnetic energy spectra $\Lambda(l)$ (see Sect. 4.1) for dynamos T2 (*solid*), T3 (*dotted*), and T5 (*dashed*). For the reversing dynamo T3 we also show a spectrum averaged over the first excursion shown in Figure 4.12. *Thin lines* drawn at a distance of half the standard deviation above and below each time average indicate the time variability

spatial resolution is relatively low. At $E = 10^{-3}$ a radial resolution of 33 grid points and a horizontal resolution which corresponds to spherical harmonic degree and order 43 was sufficient. However, even the very long simulation of dynamo $T2$ establishes no proof that this dynamo will never revers. We should thus rather be talking about reversal likelihoods instead of categorizing a numerical model as non-reversing. The conclusion that this dynamo is unlikely to revers, however, is further reinforced by the fact that the dipole tilt always stays low; there is no excursion, and even swings beyond $20°$ are rare. The onset of reversals for the lower Ekman number models of type KC02 at $E = 3 \times 10^{-4}$ may have to be refined in the future when longer runs become available.

### 4.4.3 Reversal and Excursion Properties

The multipolar dynamo state which we associated with transitional dipole configurations is characterized by two main properties: (1) an exceptionally low relative dipole strength $D$, and (2) an increased field fluctuation that also involves the weak dipole. The stronger time dependence is a result of the fact that the dipole, which varies slower than higher harmonics in the dipole dominated states, has lost its special role. Both properties have the consequence that the local magnetic signature of excursions and reversals varies strongly over the globe. This is of special significance for paleomagnetic explorations that mostly analyze data from a specific site. Local magnetic directions and intensities, inferred from rocks or sediment samples, are typically interpreted as the manifestation of a centered virtual geomagnetic dipole (VGP). The lack of spatial and temporal resolution as well as difficulties in determining absolute intensities generally prevent the construction of global paleomagnetic field models.

Following Wicht (2005) we analyze the details of reversal and excursion signatures, explore the degree of site dependence, and highlight its impact on estimates of excursion rates and event duration estimates. While Wicht (2005) analyzed model W05, we concentrate on dynamo T4 here (see Table 4.3). Roughly 17 Myr, including 21 reversals, have been simulated for this model, which suggests that the reversal rate may be somewhat lower than what it has been during the past tens of million years for Earth, 1.2 versus about 4 reversals per Myr. Rather than analyzing the full temporal and spatial information of our numerical model, we restrict ourself to resolutions that come closer to the possibilities of geomagnetic and/or paleomagnetic observations but nevertheless capture the essential dynamics. Spatial resolutions are subjected to the simulated 'mantle filter' discussed above, suppressing spherical harmonic degree $l > 14$ and damping $l = 13$ and $l + 14$. The temporal resolution corresponds to about 1300 yr.

To access the duration of an excursion or a reversal we have to define its starting and end point. Commonly, these events are considered to start when the magnetic pole, true or virtual, swings further away than a critical latitude difference $\delta\Theta_c$ from the geographic pole. They end when the magnetic pole comes closer than $\delta\Theta_c$ to either geographic pole. These kind of binary definitions have the disadvantage that incidental crossings of the critical latitude may wrongly be regarded as start or end times. We therefore use the additional criteria that excursions and reversals should be bounded by stable polarity intervals (SPIs) that last at least $T_s$. $T_s$ is typically chosen in the order of some 10 kyr, i.e. in the range of the outer core or total core dipole decay times, which amount to about 12 kyr and 29 kyr, respectively. We also neglect any excursion or SPIs shorter than $T_n = 2$ kyr to further exclude spurious insignificant behavior. Both parameters should be chosen in accordance with the magnetic field variability of the specific dynamo model.

Figure 4.15 depicts a sequence of dynamo model T4, panel (a) shows the true dipole behavior, while panels (b) and (c) show the virtual dipole tilt at two different sites. The generally larger variability of the VGPs in panels (b) and (c) reflects the additional fluctuations in the non-dipolar field contributions. Also, the number of excursions at any site is larger than the number of excursions identified when analyzing the dipole alone, and also varies significantly from site to site as demonstrated by Fig. 4.16. The mean number of excursions identified in the VGP sequences amounts to eight, while only three are found in the dipole behavior.

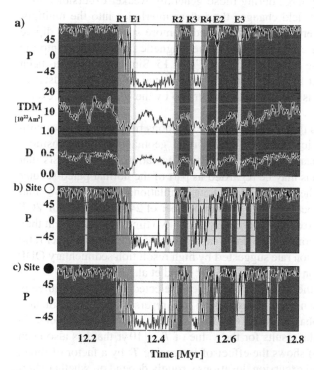

**Fig. 4.15** Panel (**a**) shows the dipole tilt P, true dipole moment TDM, and relative RMS dipole strength D during a sequence for dynamo model T4 Table 4.3. Panels (**b**), and (**c**) show the respective VGP behavior for two different sites marked in Fig. 4.16 and Fig. 4.17. *Dark grey* and white mark stable polarity epochs while mid and *light grey* mark reversals and excursions

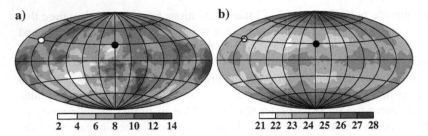

**Fig. 4.16** Global distributions of the number of locally identified excursions in VGP data (panel **a**) and ASD (panel **b**) for the sequence shown in Fig. 4.15. Earth's longitudinally averaged ASD shows a similar behavior but its amplitude is about a factor two lower (see Wicht, 2005, for a discussion on ASD). *White and black dots* mark the sites depicted in panels (**b**) and (**c**) of Fig. 4.15, respectively

Some excursions are very prominent at some sites but hardly discernable in the dipole behavior. One example is the second last excursion in panel (b) of Fig. 4.15 which can not be identified at the site depicted in panel (c). The relative dipole amplitude remains relatively large during these generally weaker excursions, they therefore rather reflect local field changes than clear interludes into the multipolar dynamo state. These interludes are associated to more severe drops in dipole strength and stronger global excursions where the magnetic pole ventures into the opposite polarity; an example is excursion E3 in Fig. 4.15. Similar to what has been found in ODP data (Lund et al., 1989), several transitional swings may cluster to form one longer event. This is, for example, illustrated by the VGP image of excursion E3 in Fig. 4.15c.

Figure 4.16 demonstrates that the number of excursions roughly correlates with the angular standard deviation (ASD) of the virtual geomagnetic pole position, which is often used as a measure for the local field variation. This suggests that the shallower local excursions may be regarded as part of the normal paleo-secular variation. When analyzing the whole 17 Myr long simulation of model T4 we find 52 excursions in the dipole behavior but a mean number of 241 excursions in VGP data. This means that dipole excursions are only about two times more frequent than reversals, while VGP excursions are ten times more likely to happen. The latter ratio agrees with the larger excursion rate suggested by high resolution sedimentary ODP records for the Brunhes chron (Acton et al., 1998; Lund et al., 2001).

A comparison between the two VGP sequences selected in Fig. 4.15 already hints at the strong site dependence of excursion and reversal duration estimates. Figure 4.17 compiles a global rendering of durations for reversal R1 depicted in Fig. 4.15. Panel (a) shows the results for the value of $T_s = 10$ yr that has also been used for Fig. 4.15, panel (b) shows the effect of increasing $T_s$ by a factor of three. The estimates of reversal and excursion durations strongly depend on whether short periods of transitional configuration are regarded as a separate excursion or not. For example, event E1 is qualified as an excursion in panels (a) and (b) of Fig. 4.15 but melts into reversal R1 at the site depicted in panel (c). When increasing $T_s$ to

a)                                                b)

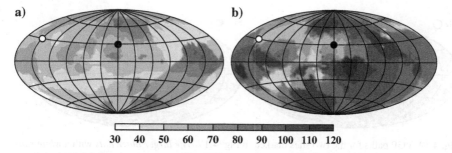

30  40  50  60  70  80  90  100 110 120

**Fig. 4.17** Global distributions of duration estimates in kyr for the first reversal shown in Fig. 4.15. A minimum stable epoch duration of $T_s = 10$ kyr and $T_s = 30$ kyr have been chosen for panels (**a**) and (**b**), respectively. $T_n$ is 2 kyr in both cases. *The dots* mark the sites depicted in Fig. 4.15

30 kyr, E1 is always regarded as part of R1 which leads to generally longer durations. However, the estimates vary by about an order of magnitude over the globe for both $T_s$ values Fig. 4.17. While we have excluded E1 from effecting the duration estimates other VGP variations remain influential.

Clement (2004) examined the latitudinal dependence of durations for the last four geomagnetic reversals and reports that the mean duration increases with latitude from about 2 kyr at the equator to about 12 kyr at 60° latitude north and south. The mean duration amounts to 7 kyr. Figure 4.18 shows that a similar latitudinal dependence emerges when averaging over several reversals in model T4. However, the durations are about an order of magnitude longer than the paleomagnetic estimates, which is more or less true for all dynamo models examined. This difference seems too large to be explained by the fact that we may include 'close by' excursions that paleomagnetists dismiss or do not capture. Singer et al. (2005) report that their duration estimate for the Matuyama-Brunhes reversal increases to about 18 kyr when they include a precursory excursion, which is still significantly shorter that our model reversals.

**Fig. 4.18** Latitudinal variation of duration estimates for reversals R1 in the sequence shown in Fig. 4.15. *Solid curve and dotted curve* are longitudinal averages of the durations shown in Fig. 4.17 for $T_s = 10$ kyr and $T_s = 30$ kyr, respectively. The *dashed curve* averages over 15 reversals in dynamo model T5 again using $T_s = 30$ kyr

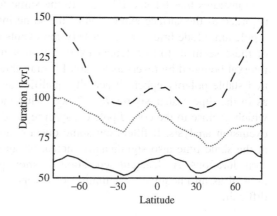

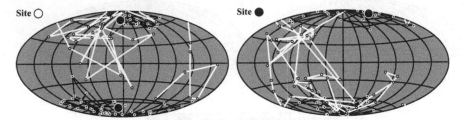

**Fig. 4.19** VGP paths for the two sites marked in Fig. 4.17. *The larger black dots* with a *white rim* mark start and end point of the depicted sequences. *White lines* mark the reversal R1, *black lines* show parts of the preceding and following stable polarity epoch

Figure 4.19 shows the VGP paths at our two example sites for a sequence that embraces R1. They illustrate that individual latitude swings can be quite fast. For example, the equatorial crossing from $45°$ degree north to $45°$ degree south takes only about 1.3 kyr for the site marked with a white dot in Fig. 4.16, which is compatible with paleomagnetic estimates. The significant differences between the two VGP sequences again highlight the important role of the multipolar field during reversals and excursions. But there are also similarities: both sites show some VGP clustering in the western hemisphere between $30°$ and $60°$ north. The VGP path for the site marked with a black dot in Fig. 4.19 stays in this area for more than 6 kyr, which is short compared to the total reversal duration, but long compared to the average VGP variability. The clustering happens at the start of the reversal and is not found at sites in the southern hemisphere. Similar clustering has been identified in paleomagnetic data (Leonhardt et al., 2002), and we will analyze the reason for this effect in Sect. 4.4.4.

Since a low $D$ value is the key spotting feature for the internal process behind reversals and excursions, it should ideally enter a duration analysis (Wicht, 2005). We decided to rely on the more commonly used characterization based on VGP latitudes since $D$ is generally not available in paleomagnetism. Figure 4.15 demonstrates that R1 and E1 fall into the same low-$D$ period and should thus be regarded as the surface expression of the same internal event. Though the relative dipole amplitude briefly recovers between events R3, R4, and E3, two or all three of these seem to form a longer excursion at some sites. The roughly 20 yr long interval bounded by reversals R3 and R4 may thus not be regarded as a firmly settled stable polarity epoch. It actually qualifies as an excursion at all sites as well as in the dipole analysis for $T_s = 30$ kyr. This indicates that this longer $T_s$ value, which is close to the core dipole decay time, leads to less ambiguous reversal and excursion analysis. It filters out some questionable shorter polarity intervals, but at the same time also significantly increases our estimates of reversal and excursion durations. A clear cut separation of short polarity intervals and excursions as well as an unequivocal definition of reversal and excursion durations remains difficult.

### 4.4.4 Transitional Field

The higher multipole components, that dominate the magnetic field during reversals or excursions, vary on time scales much shorter than the event duration. We therefore expect a generally highly time dependent field which should be subjected to a large degree of cancellation when averaged over the transitional period. The clustering of VGPs reported above (see Fig. 4.19), however, indicates that even some azimuthal features prevail. This is particularly surprising since there is no preferred longitude in the simulations. The tendency for clustering is confirmed by the VGP density shown in Fig. 4.20a, which compiles transitional VGP data from 7650 sites evenly distributed over the globe. The increased VGP density in the clustering region proves that the clustering is not a local effect encountered by a few sites only.

Figure 4.21 illustrates the variability of the magnetic energy carried by different spherical harmonic contributions during the reversal R1 depicted in Fig. 4.15. The transitional dipole low, reflected in the decrease of $l = 1$ as well as $m = 0$ contributions, constitutes the main energy variation and seems to correlate with a much weaker decrease of $l = 3$ and $l = 5$ energies. Faster variations in all contributions seem to remain comparable during stable and transitional epochs. Note that we have again chosen a time resolution of roughly 1300 yr here, which still captures the main changes and is comparable to typical resolutions in paleomagnetic data. Dipolar $l = 1$ and quadrupolar $l = 2$ energies dominate during the transitional period, axisymmetric $m = 0$ and order $m = 1$ contributions dominate the azimuthal symmetries. Order $m = 1$ maxima correlate with $l = 1$ as well as with $l = 2$ peaks, which indicates that the equatorial dipole contribution dominates only at times: this reversal is not a simple dipole tilt.

Figure 4.22 shows the radial CMB field during the times $t_1$ to $t_4$ marked by vertical lines in Fig. 4.21, we have again used a simulated mantle filter here. The solution at time $t_1$ illustrates the situation when the dipole has already decreased but has no yet tilted significantly. Note the generally weaker field indicating that not only

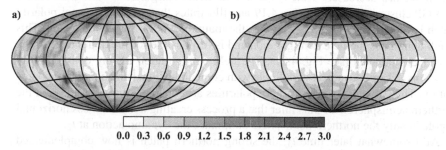

0.0  0.3  0.6  0.9  1.2  1.5  1.8  2.1  2.4  2.7  3.0

**Fig. 4.20** VGP density during reversal R1 (panel **a**) and during the whole sequence (panel **b**) shown in Fig. 4.15. The plots use transitional VGP data from 7650 sites evenly distributed over the globe. At each point we have plotted the likelihood in percent to find a VGP in a centered area of $4\pi r^2/100$ compliant with the longitude/latitude grid

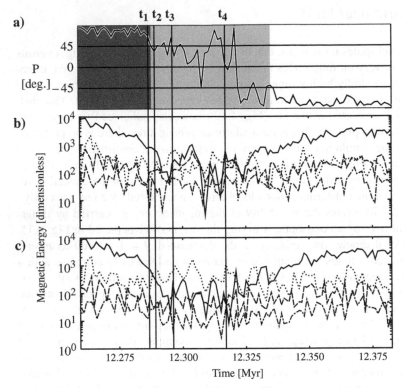

**Fig. 4.21** Magnetic CMB energy contributions of different spherical harmonic degree $l$ and order $m$ during reversal R1 shown in Fig. 4.15. Panel (**a**) shows the dipole tilt during the reversal. Panel (**b**) shows the energy contributions of degrees $l = 1$ (*solid*) $l = 2$ (*dotted*), $l = 3$ (*dashed*), and $l = 4$ (*dash-dotted*). Panel (**c**) shows the contributions of orders $m = 0$ (solid), $m = 1$ (dotted), $m = 2$ (*dashed*), and $m = 3$ (*dash-dotted*). Horizontal lines mark the times $t_1$ to $t_4$ of the radial CMB field snapshots shown in Fig. 4.22

the dipole has decreased. Time $t_2$ depicts the field configuration during the period of VGP clustering shown in Fig. 4.19 and illustrates that a strong normal polarity patch located in the clustering region is the main reason for this phenomenon. This patch dominates the field configuration and is responsible for the dominance of the $m = 1$ components (see Fig. 4.21). The field configuration mainly consists of an $(l = 2, m = 2)$ mode and a tilted dipole with comparable axial and equatorial contribution. The strongest patches of both polarities are found at similar latitudes in the northern hemisphere. We thus infer that a process creating a more or less horizontal dipole in only the northern hemisphere dominates the dynamo action at $t_2$.

At a somewhat later time $t_3$, the strong northern patch is now complemented by a southern counterpart of also normal polarity. The amplitude of the equatorial dipole component has decreased, while the energy of the axial dipole and also the $(l = 2, m = 1)$ component have increased. Finally, time $t_4$ illustrates a situation where the dynamo action is concentrated in the southern hemisphere, again producing a more or less horizontal dipole here.

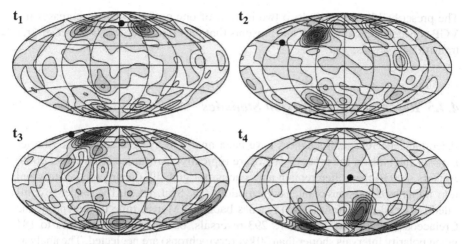

**Fig. 4.22** Snapshots of the radial magnetic field at the core–mantle boundary at the times marked in Fig. 4.21. *Black dots* show the location of the magnetic north pole

Similar features, emerging and ceasing on time scales of some kyr, dominate the field during the transitional period. The main magnetic features mostly concentrate at mid to higher latitudes, but the exact location varies with time. The time averaged field therefore remains low, and the dynamo certainly lacks the azimuthal and the north/south coherence that is responsible for the strong dipole contribution during stable polarity epochs.

Though strongly varying in amplitude, the equatorial dipole is persistent and strong enough, so that the VGP density is somewhat higher than average in a preferred latitudinal band (see Fig. 4.20). This is reminiscent of a similar effect reported by some paleomagnetic analysis (see Love, 1998, and references in there). However, the location of the preferred bands change over time in our model, and the effect averages out over several transitional events as demonstrated in Fig. 4.20b. Also, paleomagnetic data suggest that VGPs cluster at two preferred bands along American and Asian longitudes (see Love, 1998, and references therein). These bands correlate with the Pacific rim where seismic explorations find lower than average seismic velocities in the lowermost mantle. Kutzner and Christensen (2004) identify lower (higher) seismic velocities as colder (hotter) lower mantle regions where more (less) heat is allowed to escape the core. This interpretation translates lower-mantle tomographic features into a map of CMB heat flux. When using this map as a boundary condition for their dynamo simulation, Kutzner and Christensen (2004) indeed recovered the preferred VGP longitudes suggested by paleomagnetic studies. The reason is the somewhat increased convective motion below the CMB in these regions that leads to a more frequent appearance of either polarity flux patches. Upwellings as well as downwellings contribute here via field expulsion and advective field concentration respectively. The mechanism that leads to the bimodal latitudinal VGP concentration is thus in principal similar to the effect responsible for the VGP clustering described above in that it relies on local relatively short lived patches.

The prescribed heat flux leads to two instead of one preferred band, increases the VGP concentration amplitude, and prevents that the effect averages out over several transitional periods.

### 4.4.5 Reversal and Excursion Statistics

Paleomagnetic reversal sequences have been analyzed by several authors. Shorter polarity intervals in the order of 10 kyr are typically disregarded (Constable, 2000; Jonkers, 2003), they often remain ambiguous not globally documented events and may rather be excursions or intensity variations (see Sect. 4.4.3). The sequence compiled by Cande and Kent (1995) reaches back 118 Myr in time to the end of the Cretaceous superchron. It contains 293 reversals, but this number reduces to 184 when polarity intervals shorter than 30 kyr (crytochrons) are neglected. The analysis of our numerical models confirm that this minimal duration criteria helps to exclude ambiguous events. Additional data sets extend the sequence back to 160 Myr and document roughly 100 additional polarity changes (Harland et al., 1990). The relatively small number of events seems to leave room for different interpretations. A common idea is that reversals obey a poisson statistic since the scale discrepancy between polarity interval lengths and reversal durations suggests that reversals are independent events. The likelihood $P$ for a polarity interval of length $T_l$ is then proportional to $\exp(-\varepsilon T_l)$, where $\varepsilon$ is the underlying Poisson rate.

Rate $\varepsilon$ may have been constant during the past 40 Myr with roughly four reversals per Myr, decreases when going further back in time, and reaches $\varepsilon = 0$ during the Cretaceous supercron (Constable, 2000). A similar decrease has probably occurred previous to the supercron, albeit the rate of decrease is a matter of debate (Hulot and Gallet, 2003). The Poissonian distribution is often modified to incorporate shorter polarity intervals, resulting in a Gamma distribution (e.g., see Sect. 3. in Constable, 2000). Recently, (Jonkers, 2003) reported that shorter and longer polarity intervals tend to cluster.

The numerical costs of self-consistent dynamo models typically allow to simulate only a few reversals. This prevents a meaningful statistical analysis and, unfortunately, also complicates any conclusion about the reversal rate. Only rather simple models for larger Ekman numbers allow longer time integrations. The model presented by Wicht and Olson (2004) at $E = 10^{-2}$, on the other hand, reverses nearly oscillatory, a not very Earth-like behavior. The large Ekman number of model W05, listed in Table 4.3, results in a large scale solution that can be forwarded in time at relatively little numerical costs. The model nevertheless has several Earth-like characteristics, including a suitable overall flow and magnetic field geometry, a correct dipole strength, a convincing dipole dominance ($\overline{D} = 0.4$), a clear separation of stable and transitional periods ($\tau_T = 0.06$), and a stochastic reversal behavior that seems Earth-like at first sight (Wicht, 2005). Less convincing are the facts that the magnetic Reynolds number is about a factor five too small, and that the local Rossby number seems too large when compared with other reversing models (see

Sect. 4.4.2). Also note that the dynamo does not follow the regime separation shown in Fig. 4.10: we would rather expect a multipolar solution at a local Rossby number of $Ro_\ell = 0.22$.

The low numerical resolution required for this model allowed to simulate an equivalent of 160 Myr that were subjected to a statistical analysis, and we present the first results here. Note that we have always used the magnetic diffusion time of the outer core to put numbers on the dimensionless time in our models. The low value of $Rm = 92$ suggests that the ratio of the flow (overturn) time scale to the magnetic diffusion time scale is five times too low. Since the flow time scale seems more relevant for many dynamical features including reversal, the times would have to be divided by a factor of roughly 5 in order to make them comparable to Earth (Wicht, 2005). Consequently, we have increased parameters $T_s$ and $T_n$ for the classification of reversals and excursions to 90 kyr and 5 kyr, respectively.

Figure 4.23 shows dipole tilt, true dipole moment, and relative dipole strength $D$ during 15 Myr in the dynamo model. The sequence illustrates the variability in reversal rate and also once more demonstrates the difficulties in separating excursions and stable polarity intervals (see Sect. 4.4.3). For example, the dipole never really settles during the fist inverse (white background color) interval depicted, which should thus possibly be qualified as an excursion.

Figure 4.24a shows estimates of the reversal rate versus time. The mean reversal rate $\overline{R}$, simply based on the number of reversals and the length of the interval, amounts to about one reversals per Myr: $\overline{R} = 1.03$. We demonstrate the time variability of the reversal likelihood by using sliding Gauss-kernel windows of different half-widths. The reversal rate $R(t)$ in events per Myr is given by

$$R(t) = 10^{-6} \sum_i \frac{1}{(2\pi)^{1/2} \sigma} \exp\left[-\frac{1}{2}\left(\frac{t-t_i}{\sigma}\right)^2\right], \quad (4.17)$$

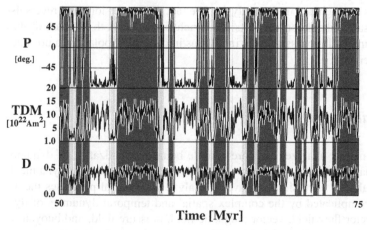

**Fig. 4.23** Dipole tilt P, true dipole moment, and relative RMS dipole strength D at the CMB for a dynamo at $E = 2 \times 10^{-2}$, $Ra = 300$, $P = 1$, and $Pm = 10$

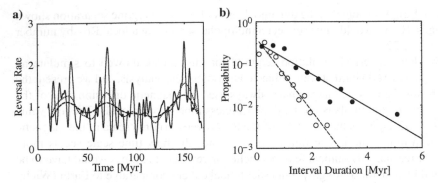

**Fig. 4.24** Reversal rate analysis for 160 Myr of the simple dynamo model W05 at $E = 10^{-2}$. Panel
(**a**) shows sliding Gauss-kernel estimates of the reversal rate for a half width of $\sigma = \overline{T}_I$ (*solid*),
$\sigma = 5\overline{T}_I$ (*dotted*) and $\sigma = 10\overline{T}_I$ (*dashed*). The mean stable polarity interval duration $\overline{T}_I$ is 1 Myr.
Panel (**b**) shows the histogram of intervals between reversals (*solid symbols*) and intervals between
reversals as well as excursions (*open symbols*). The solid and dashed line represent Poissonian
distributions fitted to both histograms, respectively

where $i$ numbers all reversals, $t_i$ is the time of a reversal (start), and $\sigma$ is the half-width. (The latter two quantities are given in years.) Estimates for $R$ vary significantly with time for a half width of $\sigma = 1$ Myr and approach the mean reversal rate when $\sigma$ is increased to 10 Myr. The same kind of analysis conducted for the paleomagnetic record shows similar results for the past 40 Myr (Constable, 2000).

Figure 4.24b shows a histogram of interval durations. Since reversals and excursions seem to have the same origin we have not only analyzed intervals between reversals but also between consecutive excursions or consecutive excursions and reversals. Both distributions seem suggestively Poissonian: the relative goodness of fit to a Poissonian distribution is about 20 % in both cases. The respective rate of the fitted distributions are $R = 0.88$ for intervals between reversals and $R = 2.00$ when excursions are included in the analysis. The fact that excursions and reversals have about the same likelihood supports the conclusions that both go back to the same origin, an internal process causing interludes into the multipolar dynamo state roughly twice per Myr.

### 4.4.6 Reversal Mechanism

So far we have mainly analyzed the core–mantle boundary realization of dynamo processes that happen deeper inside the core. To better understand the dynamical origin of reversals and excursion we have to analyze the internal processes themselves. This is complicated by the complex spatial and temporal dynamics of dynamo action: vector flow field, vector magnetic field, pressure field, and buoyancy field, all interact in a complicated fashion. The visualization of this interaction is a very demanding problem in itself.

Wicht and Olson (2004) chose to analyze a rather simple model that has many deficiencies but offeres the chance to completely untangle the reversal mechanism. Among the deficiencies of this model, that we refer to as WO04 in the following, are the non-geostrophy of the flow and the fact that the dynamo is close to being kinematic (see Sect. 4.4.2). Wicht and Olson (2004) nevertheless identified some features that may also be applicable to more realistic dynamos. A typical reversal sequence in model WO04 is illustrated in Fig. 4.25. Inverse magnetic field is mainly produced by two kinds of plumes that rise inside and outside the tangent cylinder, respectively. The role of plumes inside the tangent cylinder has already been mentioned in Sect. 4.4.2. Significantly more inverse field, however, is produce by a plume that rises outside but close to the tangent cylinder in the southern hemisphere. At some point in time, this inverse field reaches the core–mantle boundary and is then further distributed by a large scale meridional circulation. The field reversal is completed when the inverse field produced by both plume types has superseded the normal polarity field.

The duration of the reversals process at the CMB is determined by the meridional transport time. It amounts to a few thousand years for typical flow velocities inferred from geomagnetic secular variation and roughly agrees with paleomagnetic estimates (see Sect. 4.4.3). The large difference between the duration for model WO04 and model T4 (see Sect. 4.4.3) can possibly be explained by the fact that reversals in the former model consist of only one latitudinal swing but are much more complex in the latter.

Aubert et al. (2008) have developed a tool that allows to visualize the most important dynamo features even in more complex models. This tool, called DMFI for dynamical magnetic field line imager, relies on an iterative technique to highlight the regions where most of the magnetic energy is concentrated. Also, rather then using the field line density as a measure for the local field strength, DMFI focuses on a few field lines and scales their thickness according to the local magnetic energy. Aubert et al. (2008) use DMFI to analyze three different dynamos, among them

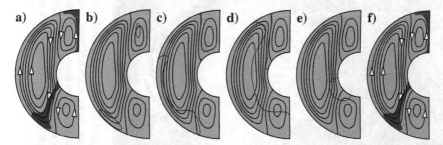

**Fig. 4.25** Sketch of a reversal sequence in model WO04 analyzed by Wicht and Olson (2004). Areas of magnetic field with the starting polarity are shown in *light red*, areas with field of opposite polarity are shown in *light blue*. The rising plumes, that are responsible for creating opposite polarity field, are illustrated in *dark red* in the first and last panel. The large scale meridional circulation, that distributes the newly created field, is shown as streamlines; *arrows* in panels (**a**) and (**f**) indicate its direction

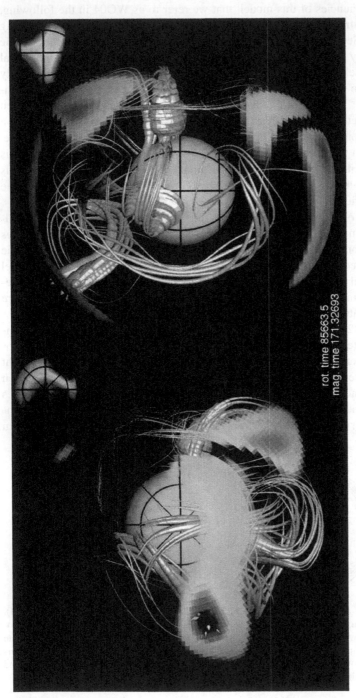

**Fig. 4.26** Magnetic field lines during a reversal visualized with the DMFI. The thickness of the field lines has been scaled with the local magnetic energy. North-polar and side view are shown at *left* and *right*, respectively. The inner core is represented by the central sphere. Color coded radial magnetic field is shown on the inner core, the outer boundary (only dominant field patches), and at a level representing Earth's surface in the upper right corner of both panels. The snapshot depicts two magnetic upwellings in the northern hemisphere that have already created enough inverse field to significantly tilt the dipole. A magnetic-anticyclone can been seen in the left hemisphere (after Aubert et al., 2008)

KC02 and W05, and identified three key dynamo features common to all models. Two of the features are the field production cycle associated with flow anticyclones (see Sect. 4.4.1), and a structure tied to flow-cyclones that also comprises the concentrated patches where these flow elements touch the CMB. Consequently, Aubert et al. (2008) named these two types of features magnetic cyclones and anticyclones, respectively. They are the main players in the process that establishes a stable dipole dominated field (see Sect. 4.4.1). A third key feature proves responsible for excursions and reversals and has been called magnetic upwellings (MUs) by Aubert et al. (2008).

Magnetic upwellings are tied to flow plumes rising inside as well as outside the tangent cylinder. They create the high-latitude inverses CMB patches mentioned in Sect. 4.4.2 and are dynamically similar to the plume features identified in model WO04. However, magnetic upwellings can amplify and concentrate field of both polarities and can therefore also create strong CMB patches of normal polarity. Figure 4.26 shows two MUs in dynamo model KC02 that have already produced enough inverse field to significantly tilt the dipole, a reversal will follow (Aubert et al., 2008). Such a large tilt is always caused by magnetic upwellings outside the tangent cylinder that produce a strong horizontal field with a significant equatorial dipole component. We speculate that the horizontal dipole features identified during the transitional period analyzed in Sect. 4.4.4 are caused by this kind of magnetic upwellings.

Magnetic upwellings rise and cease on time scales of roughly 1 kyr, but their duration as well as their amplitude vary significantly and apparently stochastically. They are essentially local features, not obeying any azimuthal or cross-equator correlation, and are thus in line with the complex unsymmetrical CMB field observed in reversing dynamo models. Weak 'local' excursions (see Sect. 4.4.3) can be caused by one MU of average amplitude. To generate strong global excursions or polarity reversals, a particularly fierce and long lasting MU is required, or several MUs have to team up to perform the job. The fact that both situations are unlikely explains the rareness of these events. While the important role of MUs in the reversal process is clearly documented by DMFI, considerably more research is required to ultimately clarify their role in the dynamo process.

## 4.5 Conclusion

We have illustrated the vitality of dynamo science with three rather diverse topics. Though fully self-consistent spherical dynamo simulations are rather successful, simplified cartesian systems remain a useful tool to study fundamental questions. Large and small Ekman number cases show important dynamical differences in the cartesian system. The observed strong influence of the Lorentz force on the flow scale is particularly remarkable since is verifies an important theoretical concept. The fact that this effect has never been observed in spherical shell simulations in puzzling. Further investigations will have to clarify whether the cartesian systems are oversimplified or whether the spherical dynamo simulations will have to go to lower Ekman numbers to capture these effects. Newly developed dynamo codes that rely on local rather than spectral methods will help in this respect. The various methods presented in Sect. 4.2 show great potential and will allow to run numerical dynamo models on modern massively parallel computers efficiently.

Our analysis of reversing dynamos have demonstrated that even the models at moderate Ekman numbers may contribute to better understand and interpret observations. We identified two different types of excursions: The more frequent local excursions, that are associated with minor latitudinal swings, show only in the VGP data of a few sites and are thus not a global phenomenon. The dipole dominance is preserved during these events. Deeper and global excursions are essentially the same type of events as complete magnetic field reversals. The key property of these events is the low relative dipole strength, they are interludes into a multipolar dynamo configuration.

The multipolar and also highly time-dependent field configuration during transitional periods goes along with a strong site dependence of VGP sequences, which in turn translates into a strong site dependence of excursion and reversal duration estimates. Despite the complexity and large time dependence of the transitional field, we were able to identify some persistent magnetic features during the transitional period and could tie these to VGP clustering that is also observed in paleomagnetic data (Leonhardt et al., 2002). Further investigations will have to clarify the significance and relevance of these features.

A particularly long computational run allowed the first meaningful statistical analysis of reversals and excursions in a numerical dynamo model. Though the reversal rate is somewhat lower than during the past million years for Earth, the statistics for paleomagnetic data and our model runs suggest a primarily Poissonian behavior. But there are also differences that will have to be investigated. For one, the model does not show the clustering of shorter or longer polarity epochs identified in the paleomagnetic sequence (Jonkers, 2003). Also, we ran our model with a thermally homogenous outer boundary though lateral thermal variations in the lower mantle can potentially influence the reversal rate (Glatzmaier et al., 1999).

The development of the new visualization tool DMFI by Julien Aubert enabled us to identify key features in the dynamo process that seem common to all models analyzed (Aubert et al., 2008). Excursions and reversals are caused by flow upwellings that locally create a strong horizontal magnetic field. This tilts the dipole and can eventually lead to a field reversal. Once again, more research is required to elucidate the detailed dynamics of these features. The considerable advancement of our dynamo knowledge combined with the advent of new numerical methods and analyzing tools promise a very productive near future for dynamo research.

# References

Acton, G., Clement, B., Lund, S., Okada, M., and Williams, T. (1998). Initial paleomagnetic results from ODP leg 172: High resolution geomagnetic field behavior for the last 1.2 Ma. *EOS*, 79:178–179.

Aubert, J., Aurnou, J., and Wicht, J. (2008). The magnetic structure of convection-driven numerical dynamos. *Geophys. J. Int.*, 172:945–956.

Blankenbach, B., Busse, F., Christensen, U., Cserepes, L., Gunkel, D., Hansen, U., Harder, H., Jarvis, G., Koch, M., Marquard, G., Moore, D., Olson, P., Schmeling, H., and Schnaubelt, T. (1989). A benchmark comparison for mantle convection codes. *Geophys. J. Int.*, 98:23–38.

Bullard, E. C. and Gellman, H. (1954). Homogeneous dynamos and terrestrial magnetism. *Philos. Trans. R. Soc. Lond.*, A. 247:213–278.

Busse, F. and Heikes, K. (1980). Convection in a rotating layer: A simple case of turbulence. *Science*, 208:173–175.

Busse, F. H. and Clever, R. M. (1979). *Nonstationary convection in a rotating system*, pp. 376–385. Recent developments in theoretical and experimental fluid mechanics: Compressible and incompressible flows. (A79-29651 11-34) Berlin, Springer-Verlag, 1979, pp. 376-385.

Cande, S. and Kent, D. (1995). Revised calibration of the geomagnetic polarity timescales for the Late Cretaceous and Cenozoic. *Geophys. Res. Lett.*, 100:6093–6095.

Cataneo, F. and Hughes, D. W. (2006). Dynamo action in a rotating convective layer. *J. Fluid. Mech.*, pp. 401–418.

Chan, K., Li, L., and Liao, X. (2006). Modelling the core convection using finite element and finite difference methods. *Phys. Earth Planet. Int.*, 157:124–138.

Chandrasekhar, S. (1961). *Hydrodynamic and Hydromagnetic Stability*. Clarendon Press.

Childress, S. and Soward, A. (1972). Convection driven hydromagnetic dynamo. *Phys. Rev. Lett.*, 29:837–839.

Christensen, U. and Aubert, J. (2006). Scaling properties of convection-driven dynamos in rotating spherical shells and applications to planetary magnetic fields. *Geophys. J. Int.*, 116:97–114.

Christensen, U. and Tilgner, A. (2004). Power requirement of the geodynamo from Ohmic losses in numerical and laboratory dynamos. *Nature*, 429:169–171.

Christensen, U. and Wicht, J. (2007). *Core Dynamics*, chapter Numerical Dynamo Simulations. Treatise on Geophysics. Elsevier.

Christensen, U. R. (2006). A deep rooted dynamo for Mercury. *Nature*, 444:1056–1058.

Christensen, U. R., Aubert, J., Busse, F. H., Cardin, P., Dormy, E., Gibbons, S., Glatzmaier, G. A., Honkura, Y., Jones, C. A., Kono, M., Matsushima, M., Sakuraba, A., Takahashi, F., Tilgner, A., Wicht, J., and Zhang, K. (2001). A numerical dynamo benchmark. *Phys. Earth Planet. Int.*, 128:25–34.

Clement, B. (2004). Dependency of the duration of geomagnetic polarity reversals on site latitude. *Nature*, 428:637–640.

Clune, T., Eliott, J., Miesch, M., Toomre, J., and Glatzmaier, G. (1999). Computational aspects of a code to study rotating turbulent convection in spherical shells. *Parallel Comp.*, 25:361–380.

Clune, T. and Knobloch, E. (1993). Pattern selection in rotating convection with experimental boundary conditions. *Phys. Rev. E*, 47:2536–2549.

Constable, C. (2000). On the rate of occurence of geomagnetic reversals. *Phys. Earth Planet. Int.*, 118:181–193.

Demircan, A., Scheel, S., and Seehafer, N. (2000). Heteroclinic behavior in rotating Rayleigh-Bénard convection. *Eur. Phys. J. B*, 13.

Dormy, E., Cardin, P., and Jault, D. (1998). MHD flow in a slightly differentially rotating spherical shell, with conducting inner core, in a dipolar magnetic field. *Earth Planet. Sci. Lett.*, 158:15–24.

Eltayeb, I. (1972). Hydromagnetic convection in a rapidly rotating fluid layer. *Proc. R. Soc. Lond. A.*, 326:229–254.

Eltayeb, I. (1975). Overstable hydromagnetic convection in a rotating fluid layer. *J. Fluid Mech.*, 71:161–179.

Fautrelle, Y. and Childress, S. (1982). Convective dynamos with intermediate and strong fields. *Geophys. Astrophys. Fluid Dyn.*, 22:235–279.

Fournier, A., Bunge, H.-P., Hollerbach, R., and Vilotte, J.-P. (2005). A Fourier-spectral element algorithm for thermal convection in rotating axisymmetric containers. *J. Comp. Phys.*, 204:462–489.

Glassmeier, K.-H., Auster, H.-U., and Motschmann, U. (2007). A feedback dynamo generating Mercury's magnetic field. *Geophys. Res. Lett.*, 34:22201–22205, doi:10.1029/2007GL031662.

Glatzmaier, G. (1984). Numerical simulation of stellar convective dynamos. 1. The model and methods. *J. Comput. Phys.*, 55:461–484.

Glatzmaier, G., Coe, R., Hongre, L., and Roberts, P. (1999). The role of the Earth's mantle in controlling the frequency of geomagnetic reversals. *Nature*, 401:885–890.

Glatzmaier, G. and Roberts, P. (1995). A three-dimensional convective dynamo solution with rotating and finitely conducting inner core and mantle. *Phys. Earth Planet. Int.*, 91:63–75.

Glatzmaier, G. and Roberts, P. (1996). An anelastic evolutionary geodynamo simulation driven by compositional and thermal convection. *Physica D*, 97:81–94.

Gubbins, D., Alfè, D., Masters, G., Price, G., and Gillan, M. (2004). Gross thermodynamics of 2-component core convection. *Geophys. J. Int.*, 157:1407–1414.

Gubbins, D., N., B. C., Gibbons, S., and J., L. J. (2000a). Kinematic dynamo action in a sphere i. effects of differential rotation and meridional circulation on solutions with axial dipole symmetry. *Proc. Roy. Soc. Lond.*, 456:1333–1353.

Gubbins, D., N., B. C., Gibbons, S., and J., L. J. (2000b). Kinematic dynamo action in a sphere ii. symmetry selection. *Proc. Roy. Soc. Lond.*, 456:1669–1683.

Gubbins, D. and Sarson, G. (1994). Geomagnetic field morphologies from a kinematic dynamo model. *Nature*, 368:51–55.

Harder, H. and Hansen, U. (2005). A finite-volume solution method for thermal convection and dynamo problems in spherical shells. *Geophys. J. Int.*, 161:522–532.

Harland, W., Armstrong, R., Cox, A., Craig, L., Smith, A., and Smith, D. (1990). *A Geological Time Scale*. Cambridge University Press, Cambridge.

Hejda, P. and Reshetnyak, M. (2003). Control volume method for the thermal dynamo problem in the sphere with the free rotating inner core. *Stud. Geophys. Geod.*, 47:147–159.

Hejda, P. and Reshetnyak, M. (2004). Control volume method for the thermal convection problem in a rotating spherical shell: test on the benchmark solution. *Stud. Geophys. Geod.*, 48:741–746.

Hulot, G. and Gallet, Y. (2003). Do superchrons occur without any paleomagnetic warning? *Earth Planet. Sci. Lett.*, 210:191–201.

Jackson, A., Jonkers, A., and Walker, M. (2000). Four centuries of geomagnetic secular variation from historical records. *Philos. Trans. R. Soc. Lond.*, A, 358:957–990.

Jones, C. (2000). Convection-driven geodynamo models. *Philos. Trans. R. Soc. Lond.*, A, 358:873–897.

Jones, C. and Roberts, P. (2000a). Convection driven dynamos in a rotating plane layer. *J. Fluid Mech.*, 404:311–343.

Jones, C. and Roberts, P. (2000b). The onset of magnetoconvection at large Prandtl number in a rotating layer II. Small magnetic diffusion. *Geophys. Astrophys. Fluid Dyn.*, 93:173–226.

Jonkers, A. (2003). Long-range dependence in the cenozoic reversal record. *Phys. Earth Planet. Int.*, 135:253–266.

Julien, K., Legg, S., McWilliams, J., and Werne, J. (1996). Rapidly rotating turbulent Rayleigh-Bénard convection. *J. Fluid Mech.*, 322:243–272.

Kageyama, A. and Watanabe, K. and Sato, T. (1993). Simulation study of a magnetohydrodynamic dynamo: Convection in a rotating shell. *Phys. Fluids*, B24:2793–2806.

Kageyama, A. and Sato, T. (1997c). Generation mechanism of a dipole field by a magnetohydrodynamic dynamo. *Phys. Rev. E*, 55:4617–4626.

Kageyama, A. and Yoshida, M. (2005). Geodynamo and mantle convection simulations on the Earth simulator using the Yin-Yang grid. *J. Phys.: Conf. Ser.*, 16:325–338.

Kono, M. and Roberts, P. (2001). Definition of the Rayleigh number for geodynamo simulation. *Phys. Earth Planet. Int.*, 128:13–24.

Krause, F. and Rädler, K. (1980). *Mean-Field Magnetohydrodynamics and Dynamo Theory*. Akademie-Verlag, Berlin.

Kuang, W. and Bloxham, J. (1999). Numerical modeling of magnetohydrodynamic convection in a rapidly rotating spherical shell: Weak and strong field dynamo action. *J. Comp. Phys.*, 153:51–81.

Kutzner, C. and Christensen, U. (2000). Effects of driving mechanisms in geodynamo models. *Geophys. Res. Lett.*, 27:29–32.

Kutzner, C. and Christensen, U. (2002). From stable dipolar to reversing numerical dynamos. *Phys. Earth Planet. Int.*, 131:29–45.

Kutzner, C. and Christensen, U. (2004). Simulated geomagnetic reversals and preferred virtual geomagnetic pole paths. *Geophys. J. Int.*, 157:1105–1118.

Leonhardt, R., Matzka, J., Hufenbecher, F., and Soffel, H. (2002). A reversal of the Earth's magnetic field recorded in mid-Miocene lava flows of Gran Canaria: Paleodirections. *J. Geophys. Res.*, 107:DOI 10.1029/2001JB000322.

Lister, J. and Buffett, B. (1995). The strength and efficiency of thermal and compositional convection n the geodynamo. *Phys. Earth Planet. Int.*, 91:17–30.

Love, J. (1998). Paleomagnetic volcanic data and geomagnetic regularity of reversals and excursions. *J. Geophys. Res.*, 103(B6):12435–12452.

Lund, S., Haskell, B., and Johnson, T. (1989). Paleomagnetic secular variation records for the last 100,000 years from deep sea sediments of the northwest Atlantic Ocean. *EOS*, 70:1073.

Lund, S., Williams, T., Acton, G., Clement, D., and Okada, M. (2001). Brunes chron magnetic field excursions recovered from leg 172 sediments. In Keigwin, L., Acton, D., and Arnold, E., editors, *Proceedings of the Ocean Drilling Program, Scientific Results*, Volume 172.

Mauersberger, P. (1956) Das Mittel der Energiedichte des geomagnetischen Hauptfeldes and der Erdoberfläche und seine säkulare Änderung. *Gerlands Beitr. Geophys.*, 65:207–215.

Matsui, H. and Okuda, H. (2005). Mhd dynamo simulation using the GeoFEM platform - verification by the dynamo benchmark test. *Int. J. Comput. Fluid Dyn.*, 19:15–22.

Maus, S., Rother, M., Stolle, C., Mai, W., Choi, S., Lühr, H., Cooke, D., and Roth, C. (2006). Third generation of the Potsdam Magnetic Model of the Earth (POMME). *Geochem. Geophys. Geosyst.*, 7:7008.

Olson, P. and Amit, H. (2006). Changes in Earth's dipole. *Naturwissenschaften.*, 93:519–542.

Olson, P. and Christensen, U. (2006). Dipole moment scaling for convection-driven planetary dynamos. *Earth Planet. Sci. Lett.*, 250:561–571.

Olson, P., Christensen, U., and Glatzmaier, G. (1999). Numerical modeling of the geodynamo: Mechanism of field generation and equilibration. *J. Geophys. Res.*, 104:10,383–10,404.

Roberts, P. and Jones, C. (2000). The onset of magnetoconvection at large Prandtl number in a rotating layer I. Finite magnetic diffusion. *Geophs. Astrophys. Fluid Dyn.*, 92:289–325.

Rotvig, J. and Jones, C. (2002). Rotating convection-driven dynamos at low ekman number. *Phys. Rev. E.*, 66:DOI:056308.

Sarson, G. and Jones, C. (1999). A convection driven geodynamo reversal model. *Phys. Earth Planet. Int.*, 111:3–20.

Singer, B., Hoffman, K., Coe, R., Brown, L., Jicha, B., Pringle, M., and Chauvin, A. (2005). Structural and temporal requirements for geomagnetic field reversals deduced from lava flows. *Nature*, 434:633–636.

Soward, A. (1974). A convection driven dynamo I: The weak field case. *Phil. Trans. R. Soc. Lond. A*, 275:611–651.

St. Pierre, M. (1993). The strong-field branch of the Childress-Soward dynamo. In Proctor, M. R. E. et al., editors, *Solar and Planetary Dynamos*, pp. 329–337.

Stanley, S. and Bloxham, J. (2004). Convective-region geometry as the cause of Uranus' and Neptune's unusual magnetic fields. *Nature*, 428:151–153.

Stanley, S., Bloxham, J., Hutchison, W., and Zuber, M. (2005). Thin shell dynamo models consistent with Mercury's weak observed magnetic field. *Earth Planet. Sci. Lett.*, 234:341–353.

Stellmach, S. and Hansen, U. (2004). Cartesian convection-driven dynamos at low Ekman number. *Phys. Rev. E*, 70:056312.

Takahashi, F. and Matsushima, M. (2006). Dipolar and non-dipolar dynamos in a thin shell geometry with implications for the magnetic field of Mercury. *Geophys. Res. Lett.*, 33:L10202.

Vorobieff, P. and Ecke, E. (2002). Turbulent rotating convection: an experimental study. *J. Fluid Mech.*, 458:191–218.

Wicht, J. (2002). Inner-core conductivity in numerical dynamo simulations. *Phys. Earth Planet. Int.*, 132:281–302.

Wicht, J. (2005). Palaeomagnetic interpretation of dynamo simulations. *GeoPhys. J. Int.*, 162:371–380.

Wicht, J. and Aubert, J. (2005). Dynamos in action. *GWDG-Bericht*, 68:49–66.

Wicht, J., Mandea, M., Takahashi, F., Christensen, U., Matsushima, M., and Langlais, B. (2007). The origin of Mercury's internal magnetic field. *Space Sci. Rev.*, 132:261–290.

Wicht, J. and Olson, P. (2004). A detailed study of the polarity reversal mechanism in a numerical dynamo model. *Geochem., Geophys,. Geosyst.*, 5:doi:10.1029/2003GC000602.

Willis, A. and Gubbins, D. (2004). Kinematic dynamo action in a sphere: Effects of periodic time-dependent flows on solutions with axial dipole symmetry. *Geophys. Astrophys. Fluid. Dyn.*, 98(6):537–554.

Zhang, K.-K. and Busse, F. (1988). Finite amplitude convection and magnetic field generation in in a rotating spherical shell. *Geophys. Astrophys. Fluid. Dyn.*, 44:33–53.

Zhang, K.-K. and Busse, F. (1989). Convection driven magnetohydrodynamic dynamos in rotating spherical shells. *Geophys. Astrophys. Fluid. Dyn.*, 49:97–116.

Zhang, K.-K. and Busse, F. (1990). Generation of magnetic fields by convection in a rotating spherical fluid shell of infinite Prandtl number. *Phys. Earth Planet. Int.*, 59:208–222.

# Chapter 5
# Effects of Geomagnetic Variations on System Earth

Joachim Vogt, Miriam Sinnhuber and May-Britt Kallenrode

## 5.1 Earth in Space

The geomagnetic field reaches out far into space where it interacts with the super-magnetosonic and magnetized solar wind to create a plasma cavity in the interplanetary medium known as the Earth's magnetosphere. Since energetic charged particles of solar and cosmic origin are affected by the magnetic field, the magnetosphere constitutes a huge particle spectrometer, and the total flux of energetic particles into the Earth's atmosphere is significantly reduced compared to the particle flux in interplanetary space at Earth's orbit. The magnetosphere acts as an efficient shield against harmful particle radiation from space.

### 5.1.1 Space Weather, Space Climate, and the Paleomagnetosphere

Spacecraft measurements in combination with ground-based observations allow to investigate the way how solar wind variations drive magnetospheric dynamics on time scales like hours or days and up to years. The fundamental physical mechanisms are reasonably well understood. Textbooks on space plasma physics (Baumjohann and Treumann, 1997; Kallenrode, 1998) discuss the key solar wind control parameters, and how solar wind induced dynamics gives rise to a wide range of so-called space weather phenomena including auroral emissions, geomagnetic storms and substorms, and atmospheric effects caused by energetic particles of solar and magnetospheric origin.

Joachim Vogt
Jacobs University Bremen School of Engineering and Science Campus Ring 1 28759 Bremen Germany

Miriam Sinnhuber
Institut für Umweltphysik Universität Bremen Otto-Hahn-Allee 1 28359 Bremen Germany

May-Britt Kallenrode
Fachbereich Physik Universität Osnabrück Barbarastrasse 7 49076 Osnabrück Germany

K.-H. Glaßmeier et al. (eds.), *Geomagnetic Field Variations*, Advances in Geophysical and Environmental Mechanics and Mathematics,
© Springer-Verlag Berlin Heidelberg 2009

The internally generated magnetic field itself is subject to variations on geological time scales. In a pioneering study on how a changing dipole moment affects the magnetospheric configuration, Siscoe and Chen (1975) coined the term paleomagnetosphere. Geomagnetic polarity transitions are well documented in paleomagnetic records (Merrill and McFadden, 1999) and can be considered the most dramatic events in the Earth's magnetic history. They must have drastically affected all aspects of the Earth's magnetosphere. The term space climate is used to comprise the phenomena that are associated with geomagnetic and solar variability on large time scales (Glassmeier et al., 2004).

## 5.1.2 Why Study the Paleomagnetosphere?

In brief, we look at paleomagnetospheric processes and space climate because (A) they are likely to affect the Earth's atmosphere and biosphere, (B) they may influence the reconstructions of the paleofield from paleomagnetic records, and (C) paleomagnetospheric research contributes to the comparative theory of planetary magnetospheres.

*Effects on the atmosphere and biosphere.* Since the magnetosphere acts as a filter for high energetic particles of solar or cosmic origin, geomagnetic variations should significantly alter particle fluxes into the atmosphere, just like solar variability does on shorter time scales. Solar energetic particles in the MeV range can contribute to the production of NOx (sum of all nitrogen containing radicals: N, NO, $NO_2$) and HOx (hydrogen containing radicals: H, OH, $HO_2$) in the middle atmosphere through dissociation and ionization of $N_2$ and $O_2$ (Crutzen et al., 1975; Porter et al., 1976; Solomon et al., 1981; Randall et al., 2001), which in turn leads to the depletion of stratospheric ozone. The formation of NOx and ozone losses were measured during several large solar particle events and can be reproduced by atmospheric chemistry models quite well (Solomon et al., 1983; Stephenson and Scourfield, 1992; Jackman et al., 2001; Rohen et al., 2005; Jackman et al., 2005a,b). Furthermore, Svensmark and Friis-Christensen (1997) suggested associations of high energetic particles and cloud coverage as well as relationships between solar activity and climate, see also the more recent work by Svensmark (2000) and Usoskin et al. (2004).

*Reconstructions of the geomagnetic field from paleomagnetic records.* Even during major magnetic storms, external current systems like the ionospheric electrojets or the ring current have been observed to contribute not more than a few percent to the present-day magnetic field at the Earth's surface. In the reconstruction of geomagnetic field parameters from paleomagnetic records, the external magnetic field is usually disregarded although, in principle, the relative contribution of the external field could be larger during weak field periods. In their pioneering work on the paleomagnetosphere, Siscoe and Chen (1975) discussed how the strength and the dynamics of the ring current are expected to behave as the Earth's dipole moment changes in time, and they estimated that the storm-time ring current may contribute as much as 42% to the total equatorial surface magnetic field during a polarity transition. If present, such a disturbance could only affect paleomagnetic records that are very well resolved in time (the temporal resolution should be of the order of days or less). Evidence for extremely rapid field changes in paleomagnetic records was found by Coe and co-workers (Coe and Prevot, 1989; Coe et al., 1995), and such rapid variations were indeed attributed to external causes by Ultré-Guérard and Achache (1995) and Jackson (1995).

*Comparative theory of planetary magnetospheres.* The importance of paleomagnetospheric research for a comparative theory of planetary magnetospheres was pointed out by Siscoe (1979). A broad range of magnetic field configurations are present at planets and other bodies in the solar system. Paleomagnetospheric studies are part of the research on comparative magnetospheres that builds primarily on observations (in-situ and remote sensing) of planetary and cometary magnetospheres but also on theoretical studies of planetary magnetospheres (Glassmeier, 1997), semi-analytical magnetospheric models (Voigt et al., 1987; Voigt and Ness, 1990), and fully three-dimensional magnetohydrodynamic (MHD) simulations (Gombosi et al., 1998; Kabin et al., 2000).

## 5.1.3 Transition Field Scenarios

In order to model paleomagnetospheric configurations, the coefficients of the internal geomagnetic field are required as input parameters. The field can change in many different ways, and we distinguish between three basic scenarios, namely, (1) variations of the dipole moment magnitude, (2) variations of the dipole moment direction, i.e., excursions of the polarity vector, and (3) non-dipolar configurations.

Since the present-day configuration is basically dipolar, the simplest type of variation to be considered is a change in the magnitude $M$ of the Earth's dipole moment **M**. The parameter $M$ is currently decreasing at the remarkable rate of about 5% per century. Put into a long-term perspective, however, this observation does not imply that we are approaching a polarity transition. Archaeomagnetic studies indicate that $M$ varied by about a factor of two over the past two thousand years (McElhinny and Senanayake, 1982). On geological timescales, the variation can be as much as one order of magnitude.

The dipole moment may also change in direction. Polarity transitions are generally preceded by several excursions of the virtual geomagnetic pole (VGP) to geographic latitudes below 45 degrees (Merrill and McFadden, 1999). Due to spatially incomplete data coverage and dating inaccuracies, however, the core field during such a polarity transition cannot be unambiguously identified from paleomagnetic records. According to one conceivable transition scenario, the dipole moment, having much reduced in magnitude, could make excursions and finally turn around. In this process, the dipole axis forms a large angle with the Earth's rotation axis for a considerable amount of time. The orientation of the geomagnetic field with respect to the solar wind flow direction changes completely in the course of one day and must lead to strong diurnal variations of the overall magnetospheric configuration.

Note that at Uranus and Neptune, the dipole axes form large angles with the rotation axis and/or the ecliptic plane normal vector.

The geomagnetic field during a polarity transition may well be essentially nondipolar. Higher-order multipoles were utilized in the interpretation of paleomagnetic records (Williams and Fuller, 1981; Clement and Kent, 1985; Clement, 1991). Large-scale geodynamo simulations (Glatzmaier and Roberts, 1996) also indicate that multipoles of higher order than the dipole are likely to dominate the Earth's core field during a polarity transition.

## 5.1.4 Earth in the Near-Space Particle Environment

The magnetosphere is bombarded by energetic charged particles from different sources. These particles can be distinguished according to their sources, energy spectra (and thus penetration depth in the atmosphere) and also by their temporal variability, as summarized in Table 5.1 and Fig. 5.1. All particle components are either protons, electrons or to a much smaller extent $\alpha$-particles. The relative contribution of heavier particles is highly variable, as is their composition.

The highest particle fluxes are carried by the solar wind, a continuous flow of coronal plasma with a density of a few particles per $cm^3$, a temperature around 1 Mio K and a flow speed of about 400 km/s, corresponding to a travel time of 4 days from the Sun to the orbit of Earth (Russell, 2001; Schwenn, 1990). The solar wind does not penetrate the magnetosphere but shapes it in its interaction with the geomagnetic field. Its density and flow speed is modulated by two factors: interplanetary coronal mass ejections (ICMEs), which are associated with the restructuring of the coronal magnetic field during periods of high solar activity, and fast solar wind streams which are associated with coronal holes. While the former occur irregularly with a higher prevalence during solar maximum, the latter are related to the solar rotation period. Both phenomena are associated with an increase in energy density and thus a compression of the geomagnetic field—the geomagnetic storm.

Magnetospheric particles (MPs) are injected into the atmosphere during such geomagnetic disturbances. The properties of precipitating MPs depend on the actual

**Table 5.1** Properties of particle populations influencing the terrestrial atmosphere: solar wind (SW), magnetospheric particles (MPs), solar energetic particles (SEPs) and galactic cosmic rays (GCRs)

|        | Source             | Energy range        | Temporal variability                   |
|--------|--------------------|---------------------|----------------------------------------|
| SW     | sun                | $\sim$1 keV         | continuous, stream structure, shocks   |
| MPs    | radiation belts    | keV to tens of MeV  | geomagnetic disturbances               |
| SEPs   | flares, shocks/CMEs | tens of keV to GeV | short term                             |
| GCRc   |                    | > some 100 MeV      | continuous, modulated with solar cycle |

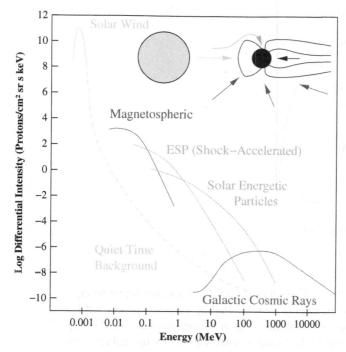

**Fig. 5.1** Spectra of energetic particle populations

properties of the radiation belt particles as summarized e.g. in Lemaire et al. (1997) and Walt (1994) and on the strength of the geomagnetic disturbance. The visible consequence of precipitating MPs is the aurora, the more energetic particles precipitate even down into the mesosphere although the bulk of MPs is already absorbed in the thermosphere. Since MPs originally have been trapped in a closed geomagnetic field configuration, they precipitate not into the polar cap but inside the polar or auroral oval surrounding the cap. As is evident from aurora observations, this oval expands equatorwards at times of strong geomagnetic activity.

Solar energetic particles (SEPs) precipitate down into the meso- and even stratosphere (Fig. 5.2); their vertical precipitation area is limited to the polar cap. The occurrence of SEP events is related to solar activity: energy previously stored in the coronal magnetic field is converted into plasma heating and acceleration, leading to flares and coronal mass ejections (CMEs). Although the details of particle acceleration and the relative contributions of flares and CMEs still are under debate, see, for example, Kallenrode (2003) and Reames (1999), their association with solar activity is not doubted and the properties of SEP events at Earth are well-established from measurements. While the relativistic SEPs arrive at Earth within tens of minutes to some hours, the CME has a travel time of 1 to 2 days. Its interaction with the magnetosphere leads to a geomagnetic disturbance and to the release of MPs into the atmosphere. As mentioned above, both sets of particles contribute to NOx production and ozone depletion (Callis et al., 1998; Crutzen et al., 1975; Porter et al.,

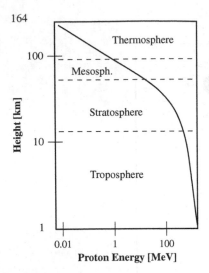

**Fig. 5.2** Penetration of energetic particle populations into different atmospheric layers

.

1976; Solomon et al., 1981; Randall et al., 2001). Since fluxes and precipitation patterns of both particle components are regulated by the geomagnetic field, a variation in atmospheric response with the variation of the geomagnetic field is expected.

The lowest fluxes but highest energies are carried by the galactic cosmic radiation (GCR) (Bieber et al., 2000; McKibben, 1987). GCRs are able to precipitate down to the lower strato- and even troposphere where they contribute to ion pair production not only during ionization but also due to secondaries created during hadronic interaction. GCRs are a continuous particle source, at its lower energetic end (up to a few GeV) modulated by the solar cycle. They precipitate globally into the atmosphere, although the lowest energies are deflected at low geomagnetic latitudes (geomagnetic cutoff). GCR precipitation might be related to cloud coverage (Svensmark and Friis-Christensen, 1997; Svensmark, 2000).

In the following we present the outcome of a collaborative effort to quantify the effects of geomagnetic variations on magnetospheric configuration and atmospheric chemistry. In Sect. 5.2 we describe various approaches to model the paleomagnetosphere ranging from scaling relations to magnetohydrodynamic simulations. Section 5.3 deals with energetic particles in the magnetosphere. The resulting particle fluxes are used to calculate ionization rates in the middle atmosphere which in turn serve as input parameters to the atmospheric model discussed in Sect. 5.4. We conclude in Sect. 5.5 with a short outlook.

## 5.2 Modeling the Paleomagnetosphere

Since the beginning of space exploration, a number of spacecraft missions visited the magnetospheres of the Earth and of other planets in our solar system. Fundamental structural elements are well understood, and general theoretical concepts can be applied to the paleomagnetosphere. As an example, Vogt and Glassmeier (2000) studied the location of trapped particle populations and the topology of the ring current in quadrupolar magnetospheres.

Most useful are scaling relations for important magnetospheric parameters like the dayside stand-off distance, the tail radius, the size of the plasmasphere, and the strength of the large-scale current systems. A number of scaling relations can be obtained for the canonical configuration where the angle between the dipole axis and the rotation axis remains small, so that the overall magnetic field topology is essentially the same as in the present-day magnetosphere. Scaling relations in the paleomagnetospheric context were derived by Siscoe and Chen (1975), Vogt and Glassmeier (2001), and Glassmeier et al. (2004).

When the dipole tilt angle is large, or non-dipolar components come into play, scaling relations are more difficult to obtain and also less useful because of the lack of reference cases. We explored basically two different routes to cope with these types of core fields, namely, a potential field approach where the external field is due to shielding currents flowing on the magnetopause (Stadelmann, 2004), and three-dimensional magnetohydrodynamic (MHD) simulations (Vogt et al., 2004; Zieger et al., 2004; Zieger et al., 2006b,a). We have been using the MHD code BATS-R-US developed at the University of Michigan, Ann Arbor, for our paleomagnetospheric studies (Gombosi et al., 1998; Kabin et al., 2000).

This section is organized in accordance with the three geomagnetic variation scenarios described in the introduction. In Sect. 5.2.1 we look at the effects of a dipole moment that changes in magnitude but remains closely aligned with the planetary rotation axis. Dipolar magnetospheres with a large tilt angle are discussed in Sect. 5.2.3. Quadrupolar paleomagnetospheres are addressed in Sect. 5.2.4 and serve as prototype examples for non-dipolar configurations.

## 5.2.1 Dipolar Configurations With Small Tilt Angles

On geological time scales the magnitude $M$ of the Earth's dipole moment can vary between 0.1 and 2 times the present value. As long as the dipole tilt angle remains small, the field topology should not change too much, and the magnetosphere can be assumed to scale in a self-similar manner. The present-day magnetosphere serves as a reference case. Figure 5.3 illustrates how the paleomagnetosphere changes in size with varying dipole moment.

## 5.2.2 Scaling of Magnetospheric Size Parameters

Siscoe and Chen (1975) were the first to derive scaling relations in the paleomagnetospheric context, namely, for the magnetopause stand-off distance as well as for the tail radius ($\propto M^{-1/3}$), for the extent of the plasmasphere ($\propto M^{-1/2}$), and for the

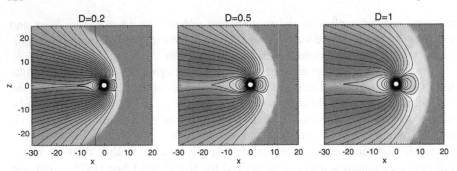

**Fig. 5.3** Simulated paleomagnetospheres with normalized dipole moments of 0.2 (*left*) and 0.5 (*middle*), and a simulation of the present-day magnetosphere (*right*) for $B_z = 10\,\text{nT}$. Magnetic field lines connected to the Earth and the thermal pressure (color-coded) are plotted in the $x - z$ plane in GSM (geocentric solar-magnetospheric) coordinates, where the $x$-axis points toward the Sun and the dipole axis is in the plane defined by the $x$ and $z$ axes; distances are given in units of $R_E$. (after Zieger et al., 2006a)

size of the polar cap ($\sin \vartheta \propto M^{-1/6}$). Vogt and Glassmeier (2001) adopted a slightly more general approach and introduced a tail radius scaling exponent that can differ from the value $1/3$. Motivated by the statistical study of spacecraft magnetopause crossings by Roelof and Sibeck (1993), we argued that this parameter should change with the north–south component $B_z$ of the interplanetary magnetic field (IMF), see also the work of Siebert (1977).

A large number of MHD simulations were carried out by Zieger et al. (2006b,a) to study how the key magnetospheric size parameters scale with the dipole moment $M$ for positive and negative values of $B_z$. To make sure that our MHD code is suited for such study, we compared the predictions of the Roelof-Sibeck bivariate function of magnetopause shape (Roelof and Sibeck, 1993) with our MHD simulation results and found them to be mutually consistent over the full validity range of the observation-based Roelof-Sibeck model. For northward IMF, the simulation results did not indicate significant deviations from the self-similar scaling. For southward IMF, i.e., negative $B_z$, the MHD simulations suggest that the magnetospheric scaling relations are indeed dependent on $B_z$ and, furthermore, that the various magnetospheric size parameters show rather different scaling behavior.

First of all, the tail radius turned out to be difficult to identify because the north–south flank distance responds differently to large negative $B_z$ values than the dawn-dusk flank distance, thus leading to an elliptically shaped magnetopause cross section. The flank distance in north–south direction is not much affected by the southward-directed IMF because the field geometry in the polar cusp regions does not favor reconnection in this case. Since the geomagnetic field is exposed to antiparallel solar wind field lines over the whole low-latitude dayside magnetopause, the dawn-dusk flanks of the magnetopause are subject to reconnection in the same way as the magnetopause at the subsolar point. In effect, the subsolar distance shows a similar scaling behavior with dipole moment as the dawn-dusk flank distance: for southward IMF, the two size parameters decrease faster with dipole moment than in the self-similar scaling model.

The left panel of Figure 5.4 displays the scaling exponent $\gamma(B_z) + 1/3$ as a function of $B_z$ in the following relation for the subsolar distance (Zieger et al., 2006a)

$$R_{ss}(B_z, M) = \alpha(B_z) \cdot R_{ss*} \cdot \left(\frac{M}{M_*}\right)^{\gamma+1/3} \tag{5.1}$$

where $R_{ss*}$ and $M_*$ denote reference values for the present-day magnetosphere at zero IMF. The function $\alpha(B_z)$ describes the decrease of $R_{ss}$ due to field erosion at the dayside magnetopause in the reference case $M = M_*$, and $\gamma(B_z)$ specifies the deviation from the ideal Chapman-Ferraro scaling due to the dipole moment dependence of the pressure balance parameter at the subsolar point of the magnetosphere. Zieger et al. (2006a) fitted a quadratic polynomial in $B_z$ to the MHD simulation results to obtain analytical functions for $\alpha$ and $\gamma$. Even for a moderately negative $B_z = -5\,\mathrm{nT}$, the function $\alpha$ gives rise to a decrease of $R_{ss}$ by about 10%. The left panel of Fig. 5.4 shows the scaling exponent $\kappa$ for the sine of the polar cap angle $\vartheta$ ($\sin\vartheta \propto M^\kappa$). For southward IMF, the polar cap size increases faster with dipole moment than in the self-similar model.

In the inner magnetosphere, one can roughly distinguish between two different plasma populations, namely, an energetic population that is ultimately fed by the solar wind during periods of magnetospheric activity, and a cold population of ionospheric origin. The energetic population gives rise to the ring current that will be addressed further below. The bulk of the cold plasma resides in the so-called plasmasphere that is rotating with the Earth. This co-rotation breaks down at the plasmapause, and the density drops by several orders of magnitude. Siscoe and Chen (1975) derived a scaling relation for the plasmapause position $R_{pp}$ assuming the transpolar potential (see below) is in its non-saturated state, and found that $R_{pp} \propto M^{1/2}$. Glassmeier et al. (2004) looked at the saturation limit and found that $R_{pp} \propto M^{-7/6}$ in this case, i.e., the plasmapause position is increasing with decreasing dipole moment. Using the paleomagnetic data SINT800 (Guyodo and Valet, 1999), we suggested that the inner paleomagnetosphere during a reversal could be dominated by co-rotation and thus resemble the present-day Jovian magnetosphere.

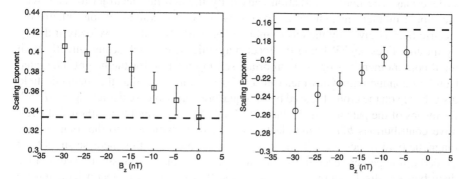

**Fig. 5.4** Dipole scaling exponents of the subsolar distance (*left*) and the polar cap size (*right*) as functions of $B_z$ (after Zieger et al., 2006a)

### 5.2.2.1 Scaling of Ionospheric and Magnetospheric Current Systems

Magnetospheric and ionospheric current systems are discussed in textbooks on space plasma physics (Baumjohann and Treumann, 1997; Kallenrode, 1998). The currents that flow on the magnetopause do not contribute significantly to the Earth's surface magnetic field. The neutral sheet current in the magnetotail region was suggested to have an effect on the geomagnetic Dst index but its contribution, if present, would be difficult to quantify in a simplified scaling model. The current systems that are thought to generate the strongest magnetic disturbance on the Earth's surface are the ring current flowing at distances of several Earth radii ($R_E$) in the equatorial region of the magnetosphere, and the auroral (or polar) electrojet (AEJ) flowing in the high-latitude ionosphere. Glassmeier et al. (2004) studied the variation of these current systems in terms of a changing dipole moment magnitude $M$, and considered also the equatorial electrojet (EEJ) flowing in the equatorial ionosphere.

Siscoe and Chen (1975) suggested that the ring current should make a significant contribution to the paleofield at the Earth's surface, and very fast field changes observed in paleomagnetic records (Coe and Prevot, 1989; Coe et al., 1995) were assumed to be caused by an enhanced ring current during geomagnetic storms (Ultré-Guérard and Achache, 1995; Jackson, 1995). Glassmeier et al. (2004) reconsidered the ring current scaling and found that the ring current index Dst should decrease with decreasing dipole moment magnitude $M$ as $Dst \propto M^{2/3}$. This view is supported by observations in Mercury's weak magnetic field which apparently does not support a ring current. Glassmeier et al. (2004) concluded that the paleo ring current cannot account for the fast variations that are occasionally observed in paleomagnetic records.

The AEJ and the EEJ are conduction currents and as such, they depend on the electric field and on the plasma conductivities. The electric field in the polar ionosphere is controlled by the level of magnetosphere-ionosphere coupling mediated by systems of field-aligned currents discussed in more detail further below. Arguing that in a weak dipolar paleomagnetosphere the transpolar potential is more likely to be driven into the saturated state, and evaluating appropriate scaling relations for the ionospheric height-integrated conductivities, Glassmeier et al. (2004) found the surface magnetic field perturbation caused by the auroral electrojet to scale with the dipole moment magnitude as $b_{AEJ} \propto M^{1/6}$. In the equatorial ionosphere, the electric field is given by the electromotive force $\mathbf{u} \times \mathbf{B}$, with the velocity $\mathbf{u}$ driven by thermal tides, and $\mathbf{B}$ being the local magnetic field which scales directly with the dipole moment. Using an appropriate scaling relation for the height-integrated Cowling conductivity, Glassmeier et al. (2004) concluded that the magnetic field disturbance on the ground caused by the equatorial electrojet scales as $b_{EEJ} \propto M^{-2/3}$. By means of the paleomagnetic data set SINT800, they found that although the relative contributions $b/B$ of the electrojet field perturbations $b$ to the total surface magnetic field $B$ increase with decreasing dipole moment, this increase should not exceed a few percent even during the most recent polarity transition. The relative disturbance is displayed in Fig. 5.5. This is still insufficient to affect paleomagnetic reconstructions.

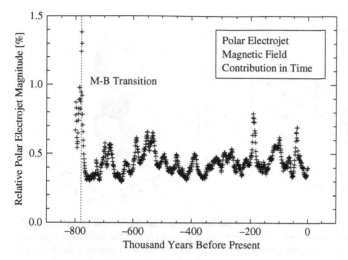

**Fig. 5.5** Estimated relative strength of the ionospheric electrojet over the past 800 000 years. *Left*: Auroral (polar) electrojet. *Right*: Equatorial electrojet (after Glassmeier et al., 2004)

A parametric model of magnetosphere-ionosphere coupling that captures the effect of transpolar potential saturation was introduced by Hill et al. (1976) and later modified by Siscoe et al. (2002). The so-called Hill-Siscoe model assumes the transpolar potential $\Phi_t$ to be

$$\Phi_t = \frac{\Phi_m \Phi_s}{\Phi_m + \Phi_s}$$

where the magnetospheric potential $\Phi_m$ is derived from a basic reconnection model and linear in the solar wind electric field. The saturation potential $\Phi_s$ is assumed to be associated with a figure-8 type system of field-aligned currents whose magnetic field counteracts dayside reconnection (see the right panel of Fig. 5.6). For $\Phi_m \gg \Phi_s$, the transpolar potential $\Phi_t$ approaches the saturation value which scales with the dipole moment as $\Phi_s \propto M^{4/3}$. Using the scaling relations for magnetospheric size parameters obtained in the MHD simulation study of Zieger et al. (2006a), we could add an IMF $B_z$ dependent correction to the Hill-Siscoe model that improved the match of the model predictions with the simulations results (see the left panel of Fig. 5.6).

## 5.2.3 Dipolar Configurations with Large Tilt Angles

Paleomagnetic records indicate that even during a polarity reversal the magnetic field does not drop to zero but maintains a finite value, so either the dipole moment has to change also in direction, or higher-order multipoles must come into play. In

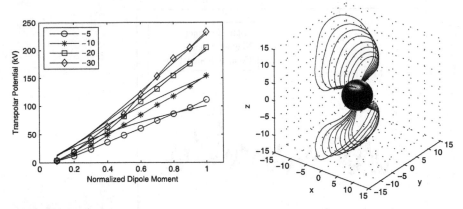

**Fig. 5.6** *Top*: Transpolar potential as a function of the normalized dipole moment for different values of $B_z$ (nT). Simulated transpolar potentials are plotted as symbols and the *solid lines* represent the corresponding curves of the corrected Hill model. *Bottom*: Closure of region 1 field-aligned currents in the present-day magnetosphere for $B_z = -5\,$nT (after Zieger et al., 2006a)

this section we consider paleomagnetospheres where the angle between the dipole axis and the Earth's rotation axis is large. The orientation of the dipole axis with respect to the solar wind flow direction changes significantly during one day, leading to global reconfigurations on time scales of several hours. We briefly refer to such a configuration as an equatorial dipolar magnetosphere. The first, mainly qualitative description was given by Saito et al. (1978).

MHD simulations of an equatorial dipolar magnetosphere were carried out by Zieger et al. (2004). The interplanetary magnetic field was assumed to follow a Parker spiral in the away sector which is one of the two most probable IMF directions of the solar wind. Figure 5.7 displays how the magnetic field line topology and large-scale magnetospheric current systems change as the dipole axis rotates in the course of one day. The dipole tilt is the angle between the dipole axis and the $z$-axis in the geocentric solar-magnetospheric coordinate system (GSM), where the $x$-axis points toward the Sun and the dipole axis is in the plane defined by the $x$ and $z$ axes. Thus in an equatorial dipolar magnetosphere the $z$ (GSM) axis points from dawn to dusk. If the dipole tilt angle is 0 degrees (first row in Fig. 5.7), there is no magnetic reconnection at the dayside magnetopause but only in the cusp regions, and the overall configuration is essentially a closed-type magnetosphere. The tail current configuration is $\Theta$-shaped with the tail magnetopause currents closing through the neutral sheet current. After six hours the dipole axis points toward the Sun (90 degrees dipole tilt, second row in Fig. 5.7), and the field line topology has changed completely. The magnetosphere opens up in response to reconnection at the nose of the magnetopause, and open field lines are convected towards the tail at the dawn side of the magnetosphere. The tail magnetopause current is detached from the neutral sheet current. Both current systems close on themselves and flow on cylindrical surfaces in opposite directions. After another six hours, the dipole tilt is 180 degrees (third row in Fig. 5.7), the perpendicular component of the IMF is

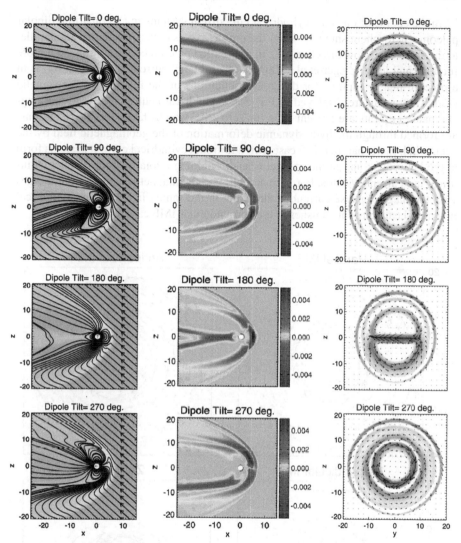

**Fig. 5.7** Diurnal variations in an equatorial dipolar paleomagnetosphere. *Left*: Magnetic field lines and pressure distribution. Center: Currents flowing across the equatorial plane. *Right*: Tail current systems. Figure composed from work of Zieger et al. (2004)

antiparallel to the geomagnetic field at the dayside magnetopause, thus yielding an open-type magnetosphere with magnetic flux transport over both poles and reconnection in the tail. The tail current system is again $\Theta$-shaped. A dipole tilt of 270 degrees (fourth row of Fig. 5.7) gives again rise to cylindrical tail current systems and marks the transition to the closed-type field line configuration at the beginning of the daily cycle.

In the course of one day, cylindrical tail current systems in an equatorial dipolar paleomagnetosphere show up when the dipole axis is either parallel or antiparallel to the solar wind flow direction, i.e., in a so-called pole-on configuration. At a more fundamental level, Zieger et al. (2004) studied the tail current system that forms in a pole-on magnetosphere for different orientations of the interplanetary magnetic field. The results are displayed in Fig. 5.8. Cylindrical neutral sheet and tail magnetopause currents are present even for zero IMF which shows that these currents are formed through the hydrodynamic deformation of the geomagnetic field by the solar wind. In the parallel IMF case, two additional cylindrical current systems form at the bow shock and in the magnetosheath, yielding a total of four distinct cylindrical tail current systems. For antiparallel IMF, the magnetosheath current merges with the magnetopause current as they flow in the same direction. The cylindrical magnetosheath current vanishes in the perpendicular IMF case.

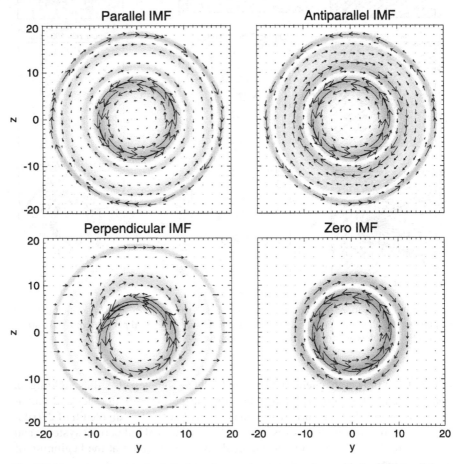

**Fig. 5.8** Current systems in the tail of a pole-on paleomagnetosphere with the IMF parallel, antiparallel, and perpendicular to the solar wind flow, and with zero IMF (after Zieger et al., 2004)

## 5.2.4 Non-dipolar Paleomagnetospheres

Reversals of the geomagnetic field are likely to produce significant non-dipolar core field components. In paleomagnetospheric research, higher-order multipoles were considered in several theoretical studies, and the case of an internal quadrupole field received special attention. Auroral zones in a quadrupolar magnetosphere were addressed by Siscoe and Crooker (1976), and the quadrupolar paleoionosphere was studied by Rishbeth (1985). Two-dimensional analytical models of quadrupolar paleomagnetospheres were constructed by several authors (Biernat et al., 1985; Leubner and Zollner, 1985; Starchenko and Shcherbakov, 1991). In a highly idealized spherically symmetric geometry, magnetopause shielding of multipoles of even higher order was studied by Willis et al. (2000).

### 5.2.4.1 The Inner Quadrupolar Paleomagnetosphere

Unlike the pure dipole case where all dipolar fields are topologically similar (i.e., the field lines of a given dipole field can be transformed into those of any other dipole field through an appropriate rotation of the coordinate system), there are different topological classes of quadrupole fields. Vogt and Glassmeier (2000) introduced the quadrupole shape parameter $\eta$ to smoothly connect two fundamental quadrupole topologies, namely, axisymmetric quadrupoles ($\eta = 0$) and those with a neutral line ($\eta = \pm 1$). The two extreme cases are included in Fig. 5.9. Topologically, any other quadrupole field can be characterized by an intermediate value of $\eta$.

Vogt and Glassmeier (2000) showed that the shape parameter $\eta$ controls the location of trapped particle populations in a quadrupolar paleomagnetosphere.

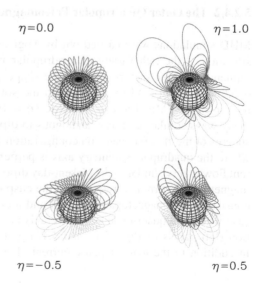

$\eta=0.0$        $\eta=1.0$

$\eta=-0.5$       $\eta=0.5$

**Fig. 5.9** Quadrupole field lines for different values of the shape parameter $\eta$. Field lines emanate from the sphere at latitudes $\beta = \pm 70°$ and $\beta = \pm 50°$ (after Vogt and Glassmeier, 2000)

**Fig. 5.10** Quadrupole
trapping center surfaces for
different values of the shape
parameter $\eta$. Superimposed
are contours of constant
magnetic field strength $B$
(*black, solid lines*), which
correspond to drift orbits of
particles with zero parallel
velocity (after Vogt and
Glassmeier, 2000)

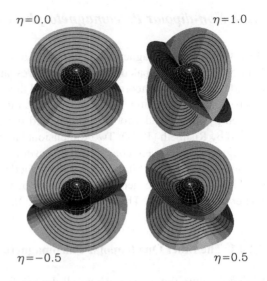

$\eta=0.0$          $\eta=1.0$

$\eta=-0.5$          $\eta=0.5$

Figure 5.10 displays the so-called "trapping center surfaces" formed by the central orbit points of trapped particles bouncing between two mirror points. The solid lines represent selected drift orbits on those surfaces. In the degenerate cases $\eta = \pm 1$ characterized by the presence of magnetic neutral lines, the drift orbits of energetic particles can easily access the atmosphere around the points where the neutral lines intersect the Earth's surface. This kind of drift orbit convergence may lead to a significant increase of energetic particle flux into the upper atmosphere in regions of very weak magnetic field.

### 5.2.4.2 The Outer Quadrupolar Paleomagnetosphere

MHD simulations were carried out by Vogt et al. (2004) to arrive at a first classification of the rich variety of quadrupolar magnetospheres. We considered the influence of three factors, namely, the shape parameter $\eta$, the orientation of the quadrupolar core field with respect to the solar wind flow direction, and the direction of the IMF. The axisymmetric ($\eta = 0$) case yields features which are in many ways similar or at least analogous to dipolar magnetospheres. In both hemispheres of the $\eta = 0$ equator-on configuration (right panels of Figs. 5.11 and 5.12) where the quadrupole symmetry axis is perpendicular to the solar wind flow, current flow is very similar to the present-day dipolar configuration: one finds a dayside magnetopause current vortex around the cusp combined with a tail current system consisting of a magnetopause current and a cross-tail current. The $\eta = 0$ pole-on case where the quadrupole symmetry axis is parallel to the solar wind flow direction (left panels of Figs. 5.11 and 5.12) yields two circular tail current systems in addition to the magnetopause current. This compares nicely with the circular

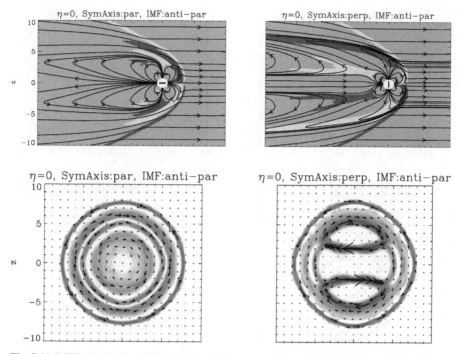

**Fig. 5.11** MHD simulations of quadrupolar paleomagnetospheres: axisymmetric quadrupolar core fields, IMF direction is antiparallel to the solar wind flow. *Left*: Pole-on configuration. *Right*: Equator-on configuration (after Vogt et al., 2004)

tail current identified by Zieger et al. (2004) in simulations of pole-on dipolar magnetospheres.

In comparison with the results of the corresponding zero IMF reference cases (see the figures 2 and 3 in Vogt et al. (2004), not shown here), purely parallel or antiparallel IMF components (see Fig. 5.11) do not affect much the field line geometry and the large-scale current systems. In particular, the bow shock current and the magnetosheath current (the latter appears to be pronounced only when there is no perpendicular IMF component) simply flow around the magnetosphere and do not interact with it. Nonzero perpendicular IMF components (see Fig. 5.12) affect the magnetospheric configuration much stronger. In both the pole-on and the equator-on case there is always one hemisphere at the dayside magnetopause where the field geometry supports reconnection. This yields large-scale transport of plasma and magnetic flux over one of the poles and subsequent tail reconnection, analogous to the classical dipolar field line merging scenario. However, in a quadrupolar magnetosphere only one hemisphere is affected which suggests that solar wind-magnetosphere coupling through reconnection, albeit more or less persistent, should happen on a smaller scale.

For $\eta = 1$ magnetospheres, the field configurations are essentially three-dimensional, and many aspects are difficult to compare with the dipolar case.

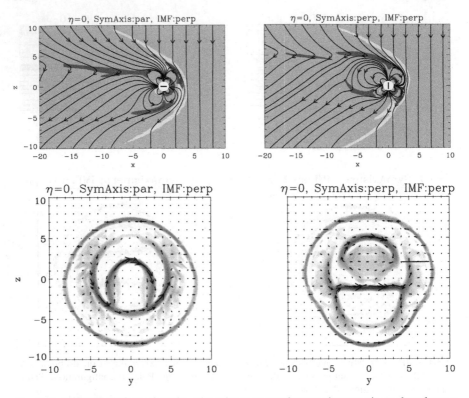

**Fig. 5.12** MHD simulations of quadrupolar paleomagnetospheres: axisymmetric quadrupolar core fields, IMF direction is perpendicular to the solar wind flow. *Left*: Pole-on configuration. *Right*: Equator-on configuration (after Vogt et al., 2004)

Vogt et al. (2004) found that tail magnetopause currents close through more complex cross-tail currents. A sample configuration is shown in Figure 5.13. As for axisymmetric quadrupoles, purely parallel or antiparallel IMF components do not change the magnetospheric configuration as much as perpendicular IMF components do. In general, weak field regions that originate from a quadrupolar core field become shielded in the process of magnetosphere formation. This suggests that the influx of high-energy particles into such regions should be less pronounced than expected from the geometry of the core field. As in the $\eta = 0$ case, perpendicular IMF components of arbitrary clock angles find antiparallel field lines of planetary origin on the dayside of $\eta = 1$ quadrupolar magnetospheres. This means that field line merging should again occur very frequently but on possibly smaller scales. In contrast to dipolar magnetospheres, paleomagnetospheric dynamics in non-dipolar configurations should thus be persistent rather than strongly dependent on the IMF orientation.

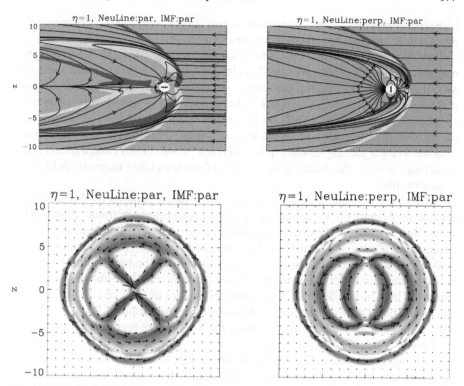

**Fig. 5.13** MHD simulations of quadrupolar paleomagnetospheres: quadrupolar core fields with neutral lines (shape parameter $\eta = 1$), IMF direction is parallel to the solar wind flow. *Left*: Neutral line aligned with the solar wind flow direction. *Right*: Neutral line perpendicular to the solar wind flow direction (after Vogt et al., 2004)

## 5.3 Energetic Particles in the Magnetosphere

Variations of the geomagnetic field affect the magnetospheric configuration and thus also the shielding efficiency of the magnetosphere against harmful particle radiation of solar and cosmic origin. Before we present analytical and numerical models that quantify the changes in particle parameters, we look at the present-day situation.

### 5.3.1 Observations in the Present-Day Magnetosphere

Temporal and spatial precipitation patterns of energetic particles into the magnetosphere are regulated by the configuration of the geomagnetic field and by superimposed geomagnetic disturbances. While the former can be regarded as a slowly varying component defining the size of the polar cap, the latter can lead to considerable changes in the geomagnetic cutoff (Leske et al., 2001). Consequently, it is worth to consider precipitation patterns in more detail.

Particle precipitation areas are defined by the geomagnetic field. Since the geomagnetic dipole axis is not only tilted but also shifted with respect to Earth's rotation axis, the description of spatial precipitation patterns of energetic particles in geographic coordinates is rather complex. A better overview is gained using polar oval coordinates (Wissing et al., 2008a,b): this means that the auroral oval is determined for each energy bin from a fit on the intensity–latitude profiles of a polar orbiting satellite during a geomagnetically quiet reference period. In this case, a fixed latitudinal ring separates the polar cap (polewards) from the closed magnetic field lines (equatorwards).

Let us start with the analysis of spatial precipitation patterns for MPs, simply because they contribute to the definition of the above reference frame. Figure 5.14 shows fluxes of 30–300 keV electrons plotted in auroral oval coordinates. The different panels correspond to different local magnetic times. In all panels the auroral oval is indicated in both hemispheres as a straight line. The blob with high fluxes at low southern latitudes is the South Atlantic Anomaly (SAA): although fluxes are extremely high inside the SAA, these particles still are trapped and do not precipitate locally but at the footpoints of the flux tubes at higher latitudes, that is inside the polar oval. Thus the particles visible inside the SAA do not represent an additional component of precipitating particles.

A noteworthy feature in Fig. 5.14 is the strong dependence of fluxes of precipitating particles inside the auroral oval in local magnetic time: electron fluxes are comparably small in the day- and evening-sectors while they are largest in the morning- and still quite large in the night-sector.

Precipitating magnetospheric protons in the hundreds of keV to MeV range also show pattern regulated by local time but opposite to that of electrons: in protons, fluxes on the morning side exceed the ones on the evening side while the opposite is true for the electrons. With increasing particle energy, the dependence on local time becomes less pronounced. The spatial precipitation pattern of MPs therefore is charge-dependent and affects particles precipitating down into the mesosphere but not the ones reaching the stratosphere.

Figure 5.14 reflects the quiet time precipitation pattern of MPs. With increasing geomagnetic activity, this pattern shifts equatorwards. Thus geomagnetic activity regulates which part of the atmosphere is affected by precipitating MPs. This effect is illustrated in Fig. 5.15 for part of the October/November 2003 event series. The middle panel shows 30–300 keV electrons plotted versus time in auroral oval coordinates, the lower panel gives fluxes of 38–53 keV electrons in interplanetary space obtained by the ACE spacecraft (Advanced Composition Explorer). During times of increased electron intensity in interplanetary space (between the two vertical blue lines), the polar cap fills with energetic electrons. A more detailed analysis shows that SEP fluxes inside the polar cap are identical to the ones measured in interplanetary space and that the polar cap is filled homogenously with particles (Bornebusch, 2005; Bornebusch et al., 2008).

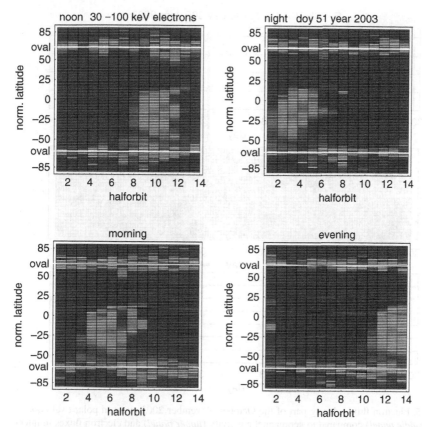

**Fig. 5.14** Dependence of 30–300 keV electron fluxes on local magnetic time plotted in auroral oval coordinates for different magnetic local times. *Upper left*: day sector, *upper right*: night sector, *lower left*: morning, *lower right*: evening (from Wissing, 2005)

The SEP event is accompanied by an interplanetary shock. While early in the SEP event fluxes of MPs are low and the sector structure is evident (circle labelled '2 sectors'), early on day 302 the shock hits the magnetosphere as evidenced by a marked increase in the Kp-index. During times of high Kp (red vertical arrows) the fluxes of MP increase markedly and the corresponding precipitation area shifts to lower latitudes. Here fluxes of MPs are much larger than SEP fluxes, although the SEP event is considered as particularly large. However, since MPs in general have a steeper spectrum than SEPs, with increasing energy the fluxes of SEPs become dominant of the MP fluxes.

The example in Fig. 5.15 demonstrates the coexistence of SEPs and MPs. This coexistence is visible in all particle energies as well as species. For atmospheric modeling thus the entire particle inventory must be considered and not only one particle species (electrons or protons) and one particle source (SEPs or MPs) at a time.

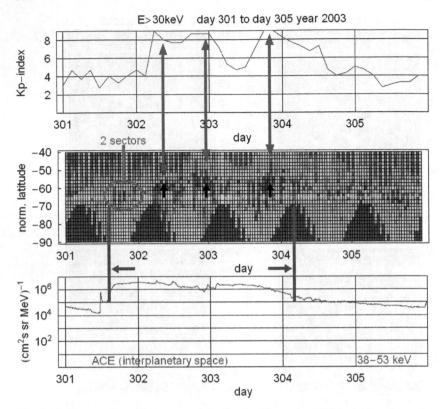

**Fig. 5.15** Electron fluxes during part of the October/November 2003 event in polar oval coordinates (*middle panel*) compared to geomagnetic activity (*upper panel*) and electron fluxes in interplanetary space (*lower panel*)(after Wissing et al., 2008a)

## 5.3.2 Analytical and Numerical Models

In a very comprehensive analysis of particle orbits in a dipole field, Størmer (1955) showed why the present-day geomagnetic field shields large parts of the Earth's surface from particles in the energy range of up to several GeV's. Numerical integration of particle orbits in more detailed model magnetic fields began in the 1960s (see, e.g., Shea et al., 1965) and first concentrated on typical galactic cosmic ray energies and particles in the internal geomagnetic (core) field only. Parametric models of the present-day magnetospheric were employed to look at daily variations of particle fluxes (Fanselow and Stone, 1972; Smart and Shea, 1972), and to determine cutoff parameters at lower energies more typical for solar energetic particles (Flueckiger and Kobel, 1990; Smart and Shea, 2001), see also the review by Smart et al. (2000).

In the context of paleomagnetospheric research, a numerical model was developed by Stadelmann (2004), and an analytical approach was chosen by Vogt et al.

(2007). In the latter study, we derived scaling relations for cutoff energies and differential particle fluxes during periods of reduced dipole moment, and quantified cutoff parameters also in axially symmetric multipole fields of higher order. Polar cap scaling was generalized to the mixed dipole–quadrupole case.

A particle tracing scheme was combined with a potential field model of the magnetospheric magnetic field by Stadelmann (2004) to form the EPOM (Energetic Particle Orbits in Magnetospheres) package. In this model, the dayside and nightside magnetopause are assumed to be spherical and cylindrical, respectively. The normal component of the magnetic field at the magnetopause can be chosen to model different levels of field line merging and solar wind-magnetosphere coupling. The internal field can be chosen as an arbitrary combination of dipolar and quadrupolar contributions. The particle tracing scheme as such solves the equations of motion of a large number of particles using the Runge-Kutta method. The upper panel of Fig. 5.16 illustrates the model geometry and the particle injection method.

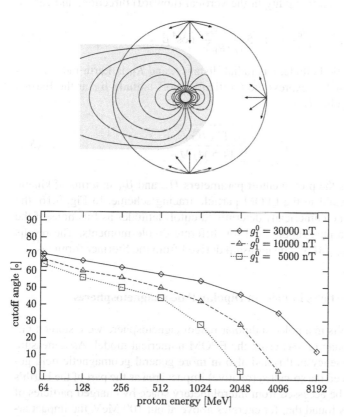

**Fig. 5.16** *Top*: Particle injection into the potential field model of the paleomagnetosphere. *Bottom*: Cutoff latitudes as functions of energy for dipolar paleomagnetospheres with different dipole moments (after Stadelmann, 2004)

### 5.3.2.1 Energetic Particles in Dipolar Paleomagnetospheres

A convenient measure of the 'reluctance' of a particle to get deflected in a given magnetic field is the rigidity $\Pi$ (momentum per charge). The relativistic relationship between momentum and kinetic energy $W$ yields for the rigidity of a proton

$$\frac{\Pi}{\text{GV}} = \sqrt{\left(0.938 + \frac{W}{\text{GeV}}\right)^2 - 0.938} \,. \tag{5.2}$$

The cutoff rigidity $\Pi_c$ gives the minimum rigidity at a given location to find an escape route out of the geomagnetic field in a given direction.

The shielding efficiency of the geomagnetic field can be characterized by cutoff parameters. For a pure dipole field, analytical formulas for cutoff rigidities were given by Størmer (1955) (see also the review by Smart and Shea, 2005) which show that cutoff rigidities $\Pi_c$ scale linearly with the dipole moment $M$. The vertical cutoff rigidity $\Pi_{cv}$ is the cutoff rigidity in the vertical (upward) direction, and can be written as

$$\frac{\Pi_{cv}}{\text{GV}} = 14.5 \frac{M/M_*}{(r/R_E)^2} \cos^4 \beta \tag{5.3}$$

where $\beta$ is geomagnetic latitude, $r$ is radial distance, and $R_E$ is Earth radius. This relationship can be used to express the (vertical) cutoff latitude $\beta_{cv}$ at the Earth's surface for a given rigidity $\Pi$ as

$$\cos \beta_{cv} = \sqrt[4]{\frac{\Pi_{cv}/\text{GV}}{14.5 M/M_*}} \,. \tag{5.4}$$

One may now express the proton cutoff parameters $\Pi_{cv}$ and $\beta_{cv}$ in terms of kinetic energy $W$, and then validate the EPOM particle tracing scheme. In Fig. 5.16, the bottom panel shows the numerically determined cutoff latitudes as functions of the kinetic energy $W$ in a pure dipole field for different dipole moments. The results follow very well the $\beta_{cv}(M, W)$ relationship derived from the Størmer formulas.

### 5.3.2.2 Energetic Particles in Pole-on Dipolar Paleomagnetospheres

Energetic particle orbits in a pole-on dipolar paleomagnetosphere were studied numerically by Stadelmann (2004) using the EPOM numerical model. As a measure for the shielding efficiency in this and also in more general geomagnetic field geometries, she introduced the so-called (relative) impact area as the part of the Earth's surface area that can be accessed from interplanetary space by charged particles of a given energy. It was found that for energies above about 100 MeV, the impact areas of a pole-on paleomagnetosphere do not differ significantly from impact areas of dipolar magnetospheres with other orientations of the dipole axis. This finding

indicates that for energies above 100 MeV the magnetic field distortions in the outer
magnetosphere (tail and magnetopause fields) have little effect on the particle orbits.

### 5.3.2.3 Energetic Particles in a Quadrupolar Paleomagnetosphere

The EPOM package (Stadelmann, 2004) was also used to study energetic particle
orbits in quadrupolar paleomagnetospheres. Figure 5.17 illustrates the output of this
model in the case of an axisymmetric quadrupolar core field, and also shows the
resulting cutoff latitudes and impact areas as functions of the particle energy. These
findings agree well with the analytical results on the basis of generalized Størmer
theory presented by Vogt et al. (2007).

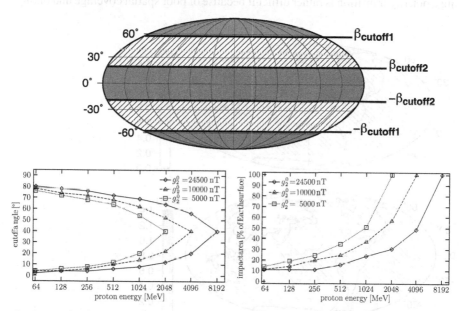

**Fig. 5.17** Energetic particle characteristics in $\eta = 0$ quadrupolar paleomagnetospheres modelled
by Stadelmann (2004) using a potential field approach. The symmetry axis is chosen to be perpendicular to the solar wind flow direction. *Upper panel*: Impact areas of 4 GeV particles. *Lower
left panel*: Cutoff latitudes as functions of energy. *Lower right panel*: Impact areas as functions of
energy (after Stadelmann, 2004)

### 5.3.2.4 Energetic Particles in a Mixed Dipole-Quadrupole
### Paleomagnetosphere

Vogt et al. (2007) generalized the polar cap scaling relations for the pure dipole
case to include contributions from an axially symmetric quadrupole field. Around

the pole on one hemisphere, the quadrupole field is directed parallel to the dipole field which leads to an enhancement of the total field strength, and the polar cap decreases in size. In the polar region on the opposite hemisphere, the two fields are antiparallel to each other, so the total field decreases in magnitude, and the polar cap increases in size. As a result, energetic particles may reach latitudes as low as 30° if the dipole moment is significantly reduced to about 10% of its present value, and the quadrupole contribution is comparable to the dipole field.

### 5.3.2.5 Impact Areas During a Simulated Field Reversal

Reconstructing the geomagnetic field coefficents from paleomagnetic records during a polarity transition is rather difficult because of poor spatial coverage and dating

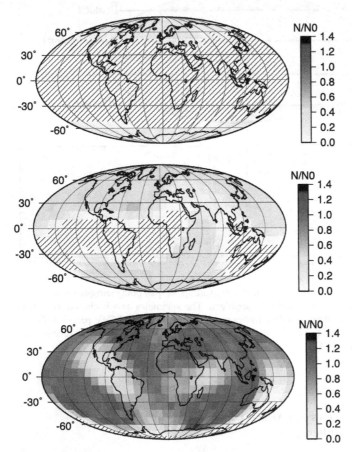

**Fig. 5.18** Regions on the globe that become accessible to 256 MeV protons before and during the simulated polarity reversal (see text). *Top*: 15,000 years before the reversal. *Middle*: 1,000 years before the reversal. *Bottom*: At the time of the reversal (after Stadelmann, 2004)

inaccuracies (Merrill and McFadden, 1999). Although computational resources are still insufficient to operate in a realistic parameter regime, numerical geodynamo simulations of the transition process can be used to identify such a field configuration at least for case studies. Stadelmann (2004) was provided with the results of the polarity transition simulation of Glatzmaier and Roberts (1995). Multipoles of higher order than the dipole dominate the magnetic energy spectrum at the time of the reversal, and protons with energies of just a few 10 MeV's can reach more than 50% of the Earth's upper atmosphere. The regions on the globe which become accessible to 256 MeV protons before and during the transition are shown in Fig. 5.18.

## 5.4 Effects on Atmospheric Composition and Chemistry

Extraterrestrial charged particles – electrons, protons and heavier ions from solar eruptions, from outside the solar system, or from the terrestrial magnetosphere – precipitating into the atmosphere can significantly alter the chemical composition of the stratosphere (10–55 km) and mesosphere (55–90 km). Particle precipitation has already been recognized as a possible source of atmospheric ozone loss in the early 1970s (Swider and Keneshea, 1973; Crutzen et al., 1975); precipitating high-energy particles ionize, decompose and excite atmospheric atoms and molecules, thus starting a chain of very fast ion chemistry reactions which transfer the chemically inert species $N_2$, $O_2$ and $H_2O$ into radicals of the NOx (N, NO, $NO_2$) (Crutzen et al., 1975; Porter et al., 1976) and HOx (H, OH, $HO_2$) (Swider and Keneshea, 1973; Solomon et al., 1981) families. Both NOx and HOx can destroy stratospheric ozone in catalytic cycles; however, while ozone destruction by HOx is restricted to the upper stratosphere and mesosphere as well as to the lowermost stratosphere, NOx is most effective for ozone loss in the mid-stratosphere (Lary, 1997).

During large particle events, ozone will be destroyed quite effectively in the mesosphere due to HOx-catalyzed cycles. However, HOx is quite short-lived in the middle atmosphere, and will recover quickly after particle fluxes have relaxed to background conditions. Equally, mesospheric ozone will recover quickly after the termination of the event. NOx on the other hand can be quite long-lived in the middle atmosphere, especially during polar winter. Thus, NOx from large particle events can be transported down into the mid- to lower stratosphere during polar winter, and can lead to long-lasting ozone losses in the so-called ozone-layer, an atmospheric layer between roughly 20–30 km altitude which contains the largest ozone densities found anywhere in the atmosphere.

A number of very large particle precipitation events from solar eruptions (so-called Solar Proton Events, SPEs) have been observed in the last three decades. Large NOx increases and mesospheric ozone losses have been observed during a number of these events. Both, NOx increase as well as mesospheric ozone loss can be reproduced by atmospheric chemistry models reasonably well (see, e.g., Jackman et al., 2001; Jackman et al., 2005a,b, and Rohen et al., 2005), indicating that the processes leading to NOx and HOx production and the subsequent mesospheric ozone loss during SPEs are understood reasonably well. Downwelling of NOx into the stratosphere was observed by the POAM satellite instrument after one large SPE in July 2000 in the southern hemisphere polar vortex (Randall et al., 2001) however, no clear signal of resultant stratospheric ozone loss could be observed due to the strong dynamical and anthropogenic variation of polar stratospheric ozone.

"Total ozone", the total amount of ozone found throughout a vertical column of the atmosphere, is the property that determines the amount of UV-B radiation absorbed by the atmosphere. Loss of total ozone was estimated by chemical models to be in the range of several percent at most, much lower than the dynamical and anthropogenic variability of ozone (Jackman et al., 2005a,b). However, it has been argued already in 1976 that this is due mainly to the shielding effect of the Earth's magnetic field, and that the impact of particle precipitation events on the total amount of ozone might be much larger during phases of greatly reduced magnetic field, like geomagnetic polarity transitions (Reid et al., 1976). It was speculated there that large solar proton events occurring during phases of geomagnetic reversal could have contributed significantly to past mass extinctions because of the large increase of harmful surface UV-B radiation that would be the consequence of the massive ozone loss. Indeed, it has been recognized that past mass extinctions often appear to happen during periods of high reversal rates (Wendler, 2004), though a number of other possible explanations by different authors have been put forward, none of which are generally accepted yet. Interestingly, many of the possible mass extinction scenarios discussed in the past also propose an increase in tropospheric/stratospheric NOx as one contributing factor of the "catastrophe" scenario. This has been discussed for bolide impacts (Prinn and Fegley, 1987), nearby supernova explosions (Ellis and Schramm, 1995; Crutzen and Bruehl, 1996), galactic gamma-ray bursts (Melott et al., 2005), transitions of the solar system through the spiral arms of our galaxy (Wendler, 2004) or an interstellar cloud (Pavlov et al., 2005), or large-scale volcanism (Toon, 1997).

Different potentially harmful consequences are discussed as the result of the NOx increase: stratospheric ozone loss and subsequent increase in surface UV-B, formation of tropospheric $HNO_3$ and subsequent acid rain, and a dimming of the atmosphere in the visible due to the increased $NO_2$ absorption, and a subsequent cooling at the surface. An increase in surface UV-B radiation (Reid et al., 1976) and the impact of an increased $NO_2$ absorption have been investigated for a past large solar proton event occurring during a magnetic field reversal (Hauglustaine and Gerard, 1990). However, the results presented in those publications are based on fairly simple model assumptions, and therefore not conclusive. Past ozone depletion events have been investigated in detail in one study (Cockell, 1999). It was found there that ozone depletion events due to bolide impacts and cosmic events are actually quite frequent, with "small" events of around 20% of global total ozone loss occurring around every couple of hundreds of years, and "large" events of more than 80% of global total ozone loss occurring around every hundred million years. Solar events were not considered in this study because they were not considered to be well enough specified yet, but were discussed as a possible contributing factor which would potentially increase the frequency of ozone-loss events. It was also stressed in this study that while it is quite difficult to find clear evidence of an "UV-B increase" event in the paleorecord, due to the high frequency of these events, many species should have evolved a certain resistance to UV-B changes; and indeed, some evidence was provided for this.

Here, we carry out model studies to investigate in detail how large solar events during phases of reduced magnetic field strength can contribute to past ozone depletion events, and what effect this could have on the surface UV-B distribution and atmospheric temperatures. The influence of energetic charged particles on atmospheric chemistry occurs in two steps:

1. the primary interaction of the precipitating energetic particle with the atmosphere is ionization.
2. the secondary interaction is the resulting chain of chemical reactions leading, for instance, to the depletion of ozone.

Consequently, the model is separated into two parts: into the derivation of atmospheric ionization based on charged particle fluxes at the top-of-atmosphere (see Sect. 5.4.1), and into the derivation of the impact of atmospheric ionization on atmospheric composition, which is carried out with a global chemistry-transport model of the middle atmosphere (see Sect. 5.4.2).

## 5.4.1 Primary Interaction: Ionization

Energetic charged particles ionize the atmosphere. The spatial ion pair production pattern can be separated into two components: the vertical ionization profile depends on energy spectra and atmospheric (density) composition while the horizontal pattern is regulated by the geomagnetic field. Note that the extend of the polar cap depends on particle energy as can be seen in particle fluxes obtained by polar orbiting spacecraft such as POES (Polar Operational Environmental Satellite, Krillke, 2006). Thus the energy spectra and ion pair production vs. height profiles depend on spatial coordinate. Modeling then occurs in three steps

1. determine energy spectra for the precipitating particles,
2. calculate ionization–height profiles from these spectra,
3. assign these ionization profiles to the corresponding vertical cells of the model atmosphere.

Conventional approaches on modeling the influence of precipitating energetic particles on atmospheric composition implicitly are based on such a separation. Some simplifications are used in this process:

- although SEPs and MPs are protons, $\alpha$s, electrons and minorities, for SEPs only protons are considered (see, e.g., Jackman et al., 2000; Jackman et al., 2001) while MPs are assumed to be electrons only (e.g Callis et al., 1998).

- following Bethe (1950) the interaction between the particles and the atmosphere is described by a continuous energy loss model.
- SEPs are assumed to precipitate into the polar cap only while MPs precipitate into the auroral oval only.
- the magnetosphere is assumed to be static although it is well-known that the geomagnetic cut-off decreases with increasing geomagnetic activity (Leske et al., 2001).
- the influence of energetic particles on the atmosphere is discussed in terms of individual events rather than as a continuous process with influences that might accumulate on longer time scales.

During the remaining part of this subsection, we will discuss the limitations of these assumptions.

### 5.4.1.1 Interaction with the Atmosphere

The standard approach in modeling the interaction of energetic charged particles with the atmosphere below about 100 km is based on Bethe–Bloch: particle inter-action is ionization only and the secondaries are not tracked. Thus all energy lost during the ionization process is deposited locally. Although mathematically sim-ple, the Bethe–Bloch formalism has three disadvantages: (A) at energies in the GeV range, hadronic interaction as an additional mechanism of energy loss sets in. This is not only important for the production of cosmogenic nuclides but also locally leads to a total energy loss of the primary and a large number of sec-ondaries which in turn interact with the atmosphere. (B) During their interaction with matter, incident electrons experience multiple-scattering. Thus their path is not a straight line but wiggles through the medium. (C) Secondary particles are not tracked and thus their ability to distribute the energy loss of the primary during one interaction over a broader spatial range is not considered. While (A) is rele-vant for the highest energies only, (B) and (C) are important at typical SEP and MP energies.

All three problems can be circumvented using a Monte Carlo simulation of the particle interaction with the atmosphere. Although Monte Carlo simulations are quite common in nuclear physics and have also been used to model the interaction of auroral particles with the thermosphere (Solomon, 2001), the interaction between precipitating particles and the meso- or stratosphere always has been modeled based on Bethe–Bloch or some approximation on it.

The Bethe–Bloch approach certainly is valid for protons and heavier nuclei with energies up to a few GeV/nucl: the primary particle still follows a straight path and the secondaries all are electrons with rather low energies. Thus the energy of the sec-ondary is deposited rather close to the primary interaction. Consequently, a Monte Carlo simulation using the full repertoire of electromagnetic interactions such as multiple scattering, Compton-scattering, ionization, photo electric effect, gamma conversion, annihilation, pair production, and production of bremsstrahlung, within

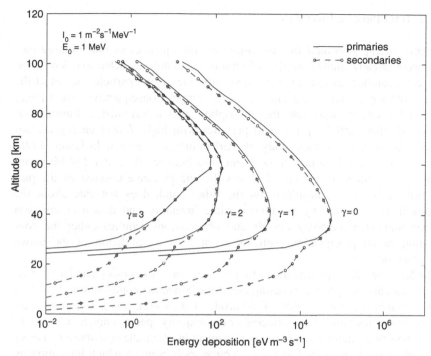

**Fig. 5.19** Energy losses for 1–50 MeV electrons in the atmosphere for different energy spectra $I(E) \sim E^{-\gamma}$ with (*symbols*) and without (*lines*) consideration of secondaries (after Schröter et al., 2006)

a reasonable spatial resolution of a few kilometers does not show any differences compared to the Bethe–Bloch solution (Schröter et al., 2006).

This is not the case for electrons: in Fig. 5.19 the solid lines give the energy deposition, which is proportional to the ionization rate, obtained from the continuous energy loss of the primary particle using a simple Bethe–Bloch algorithm (solid line) and under consideration of the secondaries (dashed lines). Both sets of curves are for electrons with energies between 1 and 50 MeV with a power-law spectrum $I(E) \sim E^{-\gamma}$ for different values of the spectral coefficient $\gamma$ ranging from 0 to 3. The main difference between both approaches is, apart from a slightly reduced energy deposition at heights above 50 km, the energy deposition below the range of the primary particle, that is below about 25 km: since the primaries are electrons, they produce Bremsstrahlung in addition to ionization. While the secondaries from the ionization process are electrons which are stopped relatively close to the primary interaction site, the Bremsstrahlung photons have a rather long range in the atmosphere and are absorbed only in the denser atmosphere, that is the lower stratosphere and the troposphere. Thus, the consideration of the secondaries leads to a significant change in the ionization–height profiles compared to the simple Bethe–Bloch approach.

#### 5.4.1.2 Total Particle Inventory

The early attempts on modeling atmospheric consequences of precipitating particles were already limited by the information available to determine ionization rates: early satellite instruments measured only one or two particle species of the SEP or MP population over a limited energy range. Consequently, only a lower limit for the energy input into the atmosphere was determined. Additional input due to other particle species and particles with higher/lower energy are neglected. For instance, in their study of SEP influences Vitt and Jackman (1996) limited their modeling to protons with energies between 0.38 and 289 MeV, not for physical reasons but because that was the energy range covered by the particle instrument. The atmosphere, on the other hand, does not care about the instrument: it is ionized by the total particle inventory and thus a comparison between ionization/chemistry models and observations requires either the consideration of all precipitating particles or at least a validation of the above approximation.

The Monte–Carlo approach allows to determine energy deposition and ionization rates for all particle species, including electrons. Since the composition of SEPs is highly variable (Cane et al., 1986; Kallenrode et al., 1992), it is not clear whether all particle species should be considered or the majority species only. Schröter et al. (2006) show two examples, one proton–rich event in which the contribution of electrons can be safely neglected and one electron–rich event in which ionization by electrons exceeds the one of protons by a factor of almost 2 in the mesosphere. Figure 5.20 shows the height-integrated energy deposition for the three major particle species in SEP events for a period of seven month in 1974. Although this time period comprises a solar minimum, individual SEP events stick out as marked increases in energy deposition. During most events the energy deposition by protons exceeds that of $\alpha$s and electrons by more than an order of magnitude, however, occasionally energy deposition by electrons and/or $\alpha$s is of the same order as that of protons.

The second objection is concerned with the energy range under study. Particles at the lower energy end define the upper boundary of the simulation volume: their stopping height gives the highest altitude up to which ion pair production rates can be simulated reliably. The opposite is not true: although the highest energies also are associated with a certain stopping height this latter can not be assumed as the lowest boundary of the simulation volume because all particles with higher energies also add ionization above this boundary. Thus the correct consideration of the highest particle energies is a prerequisite for reliable modeling. As mentioned above, older approaches extend only up to particle energies of about 300 MeV, limiting the ionization to heights above about 40 km. GOES provides proton intensities up to 500 MeV, corresponding to an energy deposit down to about 22 km. The influence of higher energies can be safely neglected: their intensities are small and the ionization produced by them concurs with the ionization produced by UV absorption in the stratosphere. In addition, Quack et al. (2001) suggested from a study of the ground level event in April 2001 that protons with energies of several hundred MeV

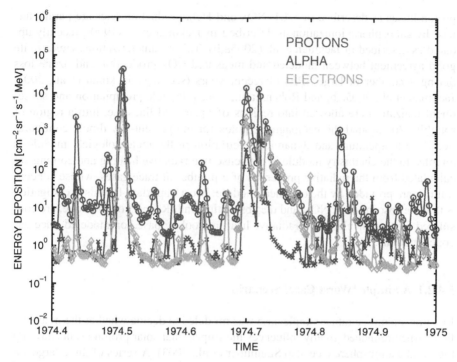

**Fig. 5.20** Energy deposition of different particle species into the atmosphere based on measurements of the IMP (Interplanetary Monitoring Platform) spacecraft (after Steinhilber, 2005)

do not contribute much to atmospheric chemistry: for a strong impact on ozone, high fluxes of SEP protons in the several tens of MeV to a few hundred MeV range are required.

## 5.4.2 Modeling Middle Atmosphere Chemistry

A global two-dimensional photochemical and transport model of the stratosphere and mesosphere has been used to investigate the impact of solar proton events in different constitutions of the geomagnetic field. The model used is a composite of the two-dimensional meteorological module THIN AIR (Kinnersley, 1996) and the chemical module of the SLIMCAT stratosphere model (Chipperfield, 1999). The meteorological module calculates temperature, pressure, and horizontal transport on isentropic surfaces, and the vertical (diabatic) transport across the isentropes. The chemical module calculates the behavior of 57 chemical species, gas-phase as well as heterogeneous. It formerly used reaction and photolysis rates from the recommendation provided by the Jet Propulsion Laboratory (JPL) for the year 2000 (Sander et al., 2000), but has recently been updated to the JPL 2006 recommendation (Sander et al., 2006). Atmospheric ionization is calculated based on measured

proton fluxes as described in 5.4.1. NOx and HOx production rates are parameterized by atmospheric ionization as described in Jackman et al. (1990), recently updated as described in Jackman et al. (2005a,b). This parameterizations yields quite
good agreement between modeled and measured NOx production and ozone loss
during a number of large SPEs of recent years (see, e.g., Jackman et al., 2001;
Jackman et al., 2005a,b, and Rohen et al., 2005). Particle precipitation and subsequent ionization are allowed into regions of open field lines, i.e. into a region of
roughly 30° around the geomagnetic poles for the present-day dipole configuration. The temperature and dynamics calculation of the meteorological module is
coupled to the chemistry module in the sense that radiative heating and cooling are
calculated from the radiative properties of a number of trace gases, whose concentrations are provided by the chemical module. Most important for this investigation
is that the absorption of $NO_2$, and the absorption and IR emission of ozone, are considered for the calculation of radiative heating/cooling based on modeled trace gas
concentrations.

### 5.4.2.1 A Simple 'Worst Case' Scenario

In a first attempt, a rather simple case was modeled to decide whether the magnetic
field structure indeed greatly influences the impact that solar proton events have on
the middle atmosphere (see also Sinnhuber et al., 2003). A series of three large solar proton events during the course of ten months was modeled, for a present day
magnetic field—meaning that particles can precipitate down into the atmosphere
only in the polar cap region > 60° geomagnetic latitude—and a scenario of a completely vanishing field – meaning that particles can precipitate into the middle atmosphere everywhere. This assumption of an atmosphere entirely exposed to energetic
particles is quite unlikely, and can be understood as an absolutely extreme case.
Ionization rates for these events were based on proton flux measurements obtained
during the October 1989 solar event, one of the largest SPEs of the past three solar
cycles. A series of three such events in a relatively short time appears to be rare,
but something similar was observed in a 400-year record based on ice-core data
for the 1890s (McCracken et al., 2001b,a). The modeled change of total ozone –
the property that determines the surface UV-B radiance – is shown in Fig. 5.21 for
the present day and the "vanishing field" scenario. A scenario with one "October
1989"-like events for a present day field is also shown. In this reference scenario
(Fig. 5.21, panel A), loss of total ozone is restricted strictly to polar regions (latitudes polewards of 50°), and is less than 2.5% in the southern hemisphere, less than
5% in the northern hemisphere. This agrees quite well with model results for the
last three solar cycles (Jackman et al., 2005a,b), and is well below the dynamical
variability of ozone (World Meteorological Organisation, 2003). Interhemispheric
differences are not a product of different proton fluxes or atmospheric ionization in
the hemispheres, but derive from the meridional circulation: as NOx is produced
well above the ozone layer even during solar proton events with high fluxes of very
high particle energies, NOx first has to be transported down into the stratospheric

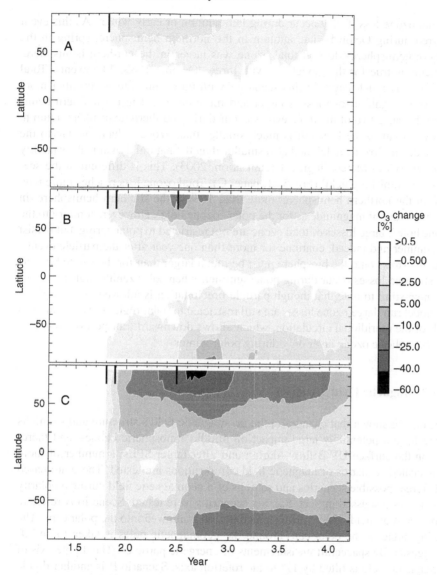

**Fig. 5.21** Modeled change of total ozone as a function of latitude for three cases: (**A**) the present day magnetic field and a "October 1989" SPE; (**B**) the present day magnetic field and three SPEs of the same magnitude as the 'October 1989' within ten months; and (**C**), the same situation as in B, but for a completely vanishing field. *Thick black lines* indicate the occurrence of the solar proton events (after Sinnhuber et al., 2003)

ozone layer to have a significant impact on total ozone. This only can happen during polar winter, when large-scale downward propagation of air-masses occurs over polar regions. Thus, as the life-time of NOx in the middle atmosphere is of the order of days to weeks decreasing with solar radiation, particle events are especially effective for total ozone loss if they occur during late autumn or early winter. As this event occurred during October—late autumn in the northern hemisphere, spring in the southern hemisphere—loss of total ozone was larger in the northern hemisphere. The same is true for the model run with three "October 1989" like events. Total ozone loss is much larger in this scenario, with maximum values reaching more than 10 %. Again, ozone loss is larger and more persistent in the northern hemisphere, because two of the three events occur during northern hemisphere autumn. However, total ozone loss still is much smaller than ozone reductions due to the anthropogenic 'ozone hole', and also smaller than the natural dynamical variability of ozone (World Meteorological Organisation, 2003). This is different in the scenario of a 'vanishing field' (Fig. 5.21, panel C). Total ozone loss reaches more than 40 % in the northern hemisphere, more than 10% in the southern hemisphere, in the same order of magnitude as for the polar 'ozone hole'. However, contrary to the "ozone hole", large losses of total ozone are not restricted to polar spring, but persist into summer and indeed, continue for more than one year after the particle events. Thus, the stress onto the biosphere must be much larger than for the ozone hole, as largest ozone losses occur during polar summer, when solar zenith angles are low. It is interesting to note that though particle precipitation is allowed everywhere in this model run, large ozone losses are still restricted to polar regions. This again is a result of the meridional circulation, which allows downward transport of NOx into the stratospheric ozone layer only during polar winter.

### 5.4.2.2 Magnetic Field Scenarios

After having shown that changes in the geomagnetic field's structure and strengths indeed have a potentially large impact on middle atmospheric ozone—and therefore, on the surface UV-B flux—during and after larger SPEs, a number of more sophisticated scenarios of magnetic field compositions are tested. These are based on different possible scenarios and settings of a geomagnetic field during a polarity transition as discussed in Sect. 5.2. Six scenarios were tested: Scenario A refers to the present-day field where particle precipitation is allowed into the polar caps. The opening angle of the polar caps around the geomagnetic poles is taken to be 30° as suggested by spacecraft measurements of energetic particles. The dipole axis of the magnetic field is tilted by 11° to the rotation axis. Scenario B is another dipole case with the same position of the field axes, but with the dipole strength reduced by 90%. The polar cap size of this configuration was calculated according to Siscoe and Chen (1975) to be 48° around the geomagnetic poles (see Sect. 5.2.1). The scaling relation of Siscoe and Chen (1975) poses an average over different orientations of the interplanetary magnetic field as discussed in Sect. 5.2.1. Scenario C is a dipole of the same strength as the present day case, but with a large tilt angle of 90°, i.e., with the geomagnetic poles in the equatorial plane as discussed in Sect. 5.2.3. The configuration chosen here poses an average over the daily variation of the magnetic field, and has polar caps of 30° width, which are centered in the equatorial belt. For all dipolar scenarios, particle precipitation is allowed only into the polar caps. Scenario D is an axisymmetric quadrupole as discussed in Sect. 5.2.4, with 30° wide polar caps and an area of open field lines around the equatorial plane, i.e., with an equatorial belt with a width of 10°. Particle precipitation is allowed into the polar caps and into the equatorial belt. Scenarios E and F are combined dipole–quadrupole cases as discussed in the last subsection of Sec. 5.2.4, and in more detail in Vogt et al. (2007). Here the dipolar component and an axisymmetric quadrupole field are assumed to make the same contribution to the total surface field at one of the poles, and are oppositely directed at the other pole. A statistical study by Constable and Parker (1988) suggested that a combined axisymmetric dipole-quadrupole field is a reasonable representation of an average reversal configuration. The (axisymmetric) dipole-quadrupole case yields two polar caps, one of 30°, and one much larger of 60°. Again, particle precipitation is allowed only into the polar cap areas. The larger polar cap lies in the southern hemisphere in scenario E, in the northern hemisphere in scenario F; both scenarios are equally likely. The magnetic field scenarios and their properties are listed also in Table 5.2. For all six scenarios, model runs were carried out for a time of eight years, with the three solar proton events of the 'worst case' scenario discussed in Sect. 5.4.2.1 in year one and two of the model run (see also Fig. 5.22). A model run without SPEs is also performed over the same time-span as a reference, and atmospheric changes are calculated relative to this reference run.

**Table 5.2** Magnetic field scenarios used for model runs

| Label | scenario | Position/size of open field line regions |
|-------|----------|------------------------------------------|
| A | present-day dipolar field | 30° around geomagnetic poles dipole tilt 11° |
| B | dipole field of reduced strength | 48° around geomagnetic poles dipole tilt 11° |
| C | equatorial dipolar magnetosphere | 30° around geomagnetic poles dipole tilt 90° |
| D | axisymmetric quadrupole | 30° around geographic poles equatorial belt of 10° |
| E | dipole-quadrupole of same strength | 30° around geomagnetic pole in NH 60° in SH |
| F | dipole-quadrupole of same strength | 60° around geomagnetic pole in NH 30° in SH |

### 5.4.2.3 Formation of Anorganic Nitrogen, Ozone Loss and the Role of the Meridional Circulation

In Fig. 5.22, the ionization rate input into the model run, the formation of total anorganic nitrogen ($NO_y = NO_x + HNO_3 + 2N_2O_5 + ClONO_2 + BrONO_2 + HNO_4$) due to the three large SPEs, and the subsequent ozone loss are shown for a latitude of 75.8°N, exemplarily for model run 'A', the present-day field. Formation of large

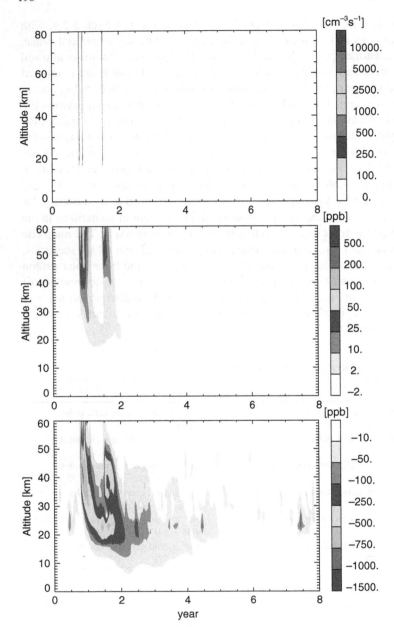

**Fig. 5.22** Ion pair production rates within the polar caps, as used as input for the model studies (*upper panel*); modeled change of NOy (*mid panel*) and ozone (*bottom panel*) at 75.8°N due to the three SPEs, relative to a model run without SPEs, for a present-day magnetic field (scenario "A")(from Winkler et al., 2008a)

amounts of NOy—the background value being in the order of magnitude of several parts per billion (ppb)—can be observed directly during the events of enhanced ionization rates at altitude above 40 km. As NOy is fairly long-lived in polar latitudes especially during winter, the NOy enhancement continues for a long time after the events, and is eventually transported down into the middle and lower stratosphere. Significant ozone losses are observed during the events above 35 km; those ozone losses are due to enhancements of HOx during the event, which is very effective for ozone loss in the upper stratosphere and mesosphere. However, HOx is also fairly short-lived, and HOx- and ozone-values recover quickly to background conditions after the event in these altitudes. As soon as NOy is transported down into the middle to lower stratosphere where NOx is most effective for catalytic ozone loss, moderate ozone losses in those altitude regions follow, as observed in Fig. 5.22. Those ozone losses continue for a long time, for months and even years after the particle event; indeed, in this example of three very large consecutive events, ozone in the lower stratosphere (around 20–30 km altitude) recovers only in year 5 of the model run, nearly four years after the first particle event. As this is the altitude range of the stratospheric ozone layer, those ozone losses will not only affect ozone in these altitude regions, but also the total ozone amount, and therefore, the surface UV-B radiation.

Loss of total ozone for this series of SPEs is shown for all magnetic field scenarios in Fig. 5.23. Maximum ozone loss ranges from less than 8% for scenario C, significantly less than in the present day scenario A, to 20–40 % in scenario F, the combined dipole-quadrupole. Also, the time until the ozone layer recovers back to values of less than 2% ozone loss varies from around 2 years for scenario C to around 4 years for scenarios B, D, E, and F. However, there are also some similarities between total ozone loss in all model scenarios, for all scenarios, the largest ozone losses are restricted mainly to polar regions, ozone loss is larger in the northern hemisphere, and ozone loss continues for a longer time in the southern hemisphere. This is again due to the mean meridional circulation of the middle atmosphere, which allows downward transport of NOx into the "ozone layer" only during polar winter. The first two of the three events occur during northern hemisphere late autumn and early winter, the third event occurs during northern hemisphere late summer; thus, the impact is larger in the northern hemisphere, but lasts for a longer time in the southern hemisphere because the last event occurs during southern hemisphere winter. For the same reason, the largest ozone losses occur in polar regions, as this is where NOx is transported down into the stratosphere. Also, ozone losses can continue for a longer time in polar regions for two reasons: (I) the main loss process of NOx and NOy in the middle atmosphere is the reaction

$$NO + N \rightarrow N_2 + O. \tag{5.5}$$

This reaction depends strongly on the availability of sunlight, as the main source of atomic nitrogen is photolysis of NO. Therefore, the life-time of NOx is longer in polar regions because of the weaker solar radiation. (II) Ozone is formed in the middle atmosphere by the three-body reaction of atomic and molecular oxygen:

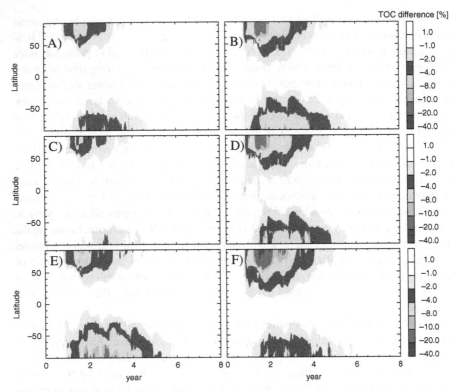

**Fig. 5.23** Modeled change of total ozone due to the three SPEs of the model time-series, relative to a model run without SPEs, for all six scenarios introduced in Table 5.2. (**A**), scenario "A", (**B**), scenario "B" etc. (from Winkler et al., 2008a)

$$O + O_2 + M \rightarrow O_3 + M .\qquad (5.6)$$

Again, atomic oxygen is formed mainly by photolysis of $O_2$, therefore this reaction depends critically on the availability of sun-light; ozone recovers more quickly in low latitudes or during polar summer than during polar autumn or winter. Thus, even for scenarios which have most or all particle precipitation in the tropics (scenario C), the largest ozone losses are observed in polar regions. However, it should be noted that some ozone loss occurs directly during the three events also in tropical regions in scenario (D); but because of the short photochemical life-time of NOx in tropical latitudes, and because of the very fast re-formation of ozone at low solar zenith angles, ozone recovers fairly quickly there. Some ozone loss is observed in mid-to low latitudes ($20° - 50°$) in the scenarios which have large polar caps in polar regions (scenarios B, E, and F); this happens some time after the particle events, and is due probably to outflow – horizontal transport – of NOx-rich and ozone-poor air from the polar stratosphere into lower latitudes, a phenomenon well known from the anthropogenic ozone hole. Also, it should be noted that the scenario which is supposed to give the most realistic representation of a magnetic field setting during a reversal, the combined dipole-quadrupole (scenarios E and F), also show the largest impact on total ozone, and thus, probably has the largest impact on the biosphere. This scenario gives significantly lower ozone losses than the simple "worst case" scenario of a completely vanishing field discussed in Sect. 5.4.2.1 for two reasons. The most fundamental reason is that the global amount of NOy produced in this simple model is much larger, as ionization was allowed everywhere instead of being restricted to areas of open field lines or polar caps. Additionally, since the model calculations of Sect. 5.4.2.1 were carried out, the process of NOx formation in the model has been changed slightly from producing NOx in the form of NO to producing a mixture of N and NO as discussed, i.e., in Jackman et al. (2005a,b). This yields slightly lower NOx production rates, as the reaction of N + NO is the major sink for NOx in the middle atmosphere, but is supposed to reproduce the mechanism of NOx formation more realistically.

### 5.4.2.4 Impact on Surface UV

The most obvious result of large losses of total ozone are changes in the surface UV-B radiation in the wavelength range 280–315 nm. Radiation of this wavelength range usually is absorbed by the ozone layer quite effectively, and every decrease of total ozone increases the surface UV-B radiation significantly. UV-B radiation can have a quite harmfull impact on the biosphere, causing damage to terrestrial plant life, single-cell organism, and aquatic ecosystems, as well as an increase in erythema and weakening of the immune system in humans (World Meteorological Organisation, 2003), thus large changes in ozone can have a potentially catastrophic impact on the biosphere. There is a nearly linear relationship between surface ozone and UV-B radiation (World Meteorological Organisation, 2003). From this, a simple linear relationship has been derived between ozone column change and erythemal weighted UV-B increase (World Meteorological Organisation, 2000). This has been used to calculate the increase of erythemal weighted UV-B for model scenario F, which has the largest impact on total ozone. The result, relative to the model run without SPEs, is shown in Fig. 5.24. A significant increase of surface UV-B radiation is observed as a result of the large SPEs in model years 1–4. The increase is largest in polar latitudes, especially in the northern hemisphere, but also reaches far into northern mid-latitudes: more than 1% of UV-B increase is observed at latitudes down to 10°N, more than 5% at latitudes down to 30°N. Maximum values of more than 20% increase are observed at latitudes north of 70°. Surface UV-B recovers to background values only 4–5 years after the first event.

### 5.4.2.5 Radiative Heating and Atmospheric Temperatures

Ozone is the key species in radiative (short-wave) heating of the stratosphere, and one of the key species in radiative (long-wave) cooling of the stratosphere; therefore, significant changes in stratospheric ozone will also impact stratospheric temperatures,

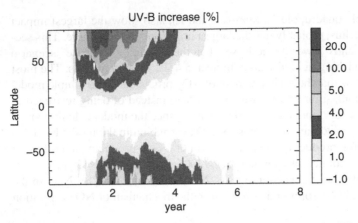

**Fig. 5.24** Modeled change of surface UV-B flux for model scenario "F" (from Winkler et al., 2008a)

and as a consequence, can also potentially impact the dynamics of the middle atmosphere, which is driven by breaking gravity waves, large-scale temperature gradients, and diabatic cooling. Also, $NO_2$ is a strong absorber in the green part of the visible spectrum, and large changes in stratospheric $NO_2$ could therefore lead to a dimming of solar light, and to a cooling of the troposphere. Both the radiative heating and cooling of the stratosphere due to changes of ozone, as well as the tropospheric cooling due to changes in $NO_2$ should be captured at least qualitatively by the models' interactive chemistry-radiation scheme. The modeled changes of atmospheric temperature from the surface up to the mesopause due to solar proton events for scenario F (the combined dipole-quadrupole) are shown in Fig. 5.25, exemplarily for a latitude of 76.8°N.

Significant changes of atmospheric temperature are observed for all three particle events. Those changes follow a strict pattern: during the events, a cooling is observed at altitudes between roughly 50–70 km; below this area of cooling, an area of atmospheric warming is observed. The reason for the cooling is the decrease of atmospheric heating due to a reduced absorption of solar light in areas of decreased ozone levels. This effect is much stronger for the second event, because it occurs during northern hemisphere summer, where more solar light is available and absorption therefore plays a larger role. The small area of warming below the cooling area is probably due to a 'self-healing' effect: because less radiation is absorbed in higher altitudes, more radiation is available in lower altitudes, which have not experienced ozone loss yet, and therefore atmospheric heating due to absorption is larger. Another area of cooling is observed around 20 km altitude starting some time before the third event, probably due to downward propagation of NOy from the first two events, and subsequent ozone loss. Changes in atmospheric temperature might in turn change the strength of the atmospheric wind patterns and atmospheric circulation; however, the significant temperature changes last for a comparatively short time of less than 1.5 years after the first particle event, thus

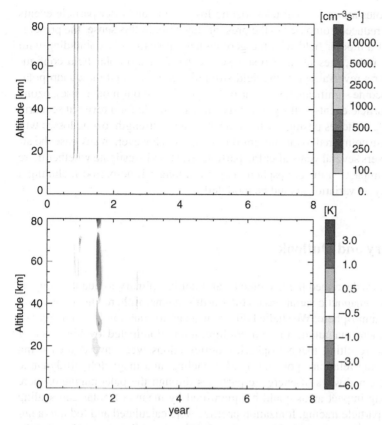

**Fig. 5.25** Modeled change of atmospheric temperature at 76.8°N for model scenario "F". Shown is the difference "F" – base; thus red contours refer to a warming, blue contours to a cooling. Ionization rates are shown in the upper panel as a reference of where particle events occurred (from Winkler et al., 2008a)

their impact on atmospheric dynamics is probably small. No significant impact on tropospheric temperatures are modeled, and thus the impact of $NO_2$ increase on atmospheric dimming and tropospheric temperatures appears to be negligible even for quite large particle events.

## 5.4.3 Conclusions

We have shown that the structure and strength of the geomagnetic field—as in the extreme case of a reversal—can have quite a large impact on atmospheric total ozone, which in turn affects the surface UV-B radiation as well as atmospheric temperatures. Largest ozone losses occur if large regions of open field lines lie within (geographic) polar regions, while in magnetic dipolar field configurations where the geomagnetic poles are in low latitudes, ozone loss during and after particle events is significantly reduced compared to the present-day case. In this sense, the present-day scenario of a dipolar field with the geomagnetic poles nearly coinciding with the geographic poles, presents a "worst-case" scenario of a dipolar field configuration. Thus, for a reversal where the field strength decreases, and the quadrupolar components become significant (as scenarios "E" and "F" of our model case), ozone losses due to particle events will significantly increase, while for a reversal scenario where the dipole switches position without decreasing in strength, ozone losses will actually decrease compared to the present-day scenario. However, in any case the atmosphere recovers several years after the particle event; to investigate whether there is a long-term impact on the atmosphere as well, a longer time-period including a realistic solar cycle variation should be modeled.

## 5.5 Summary and Outlook

In this chapter, we described the outcome of an interdisciplinary to quantify important effects of geomagnetic variations on the Earth's magnetosphere, the ionosphere, and the middle atmosphere. We studied different scenarios ranging from changes of the dipole moment magnitude to field configurations dominated by higher-order multipoles. The resulting magnetospheric configurations were investigated using analytical mapping relations, potential field modeling, and magnetohydrodynamic simulations. The pathways of energetic particles through the paleomagnetosphere and the resulting impact areas could be quantified by means of polar cap scaling and numerical particle tracing. Ionization profiles were calculated and fed into a numerical model to compute the effects on the chemistry and the composition of the paleoatmosphere, and it was shown that the magnetic field configuration can have quite a large impact on stratospheric ozone and surface UV-B radiation during and after large solar particle events.

To evaluate long-term influences of precipitating particles on atmospheric chemistry in different magnetic field configurations, we need a long-term base of precipitating particles. Continuous SEP observations, however, are available only from the early 1970s with the IMP and GOES satellites, that is a time period of a little more than three solar cycles. In addition, very large SEP events are indicated by thin nitrate layers in ice cores and have been evaluated for the last 400 years (McCracken et al., 2001a,b) – indicating that the Sun has been unusually inactive during the last decades even during solar maximum while earlier SEP fluxes have been up to almost an order of magnitude larger than the maximum fluxes observed in the satellite age.

A stochastic simulation based on event size and waiting time distributions observed by IMP and also including the ice-core events yields a kind of stochastic oscillation in cycle length and strength (Steinhilber, 2005). Results of this simulation are used in a recent study (Winkler et al., 2008a) to model atmospheric composition changes of different magnetic field scenarios over time-periods of several centuries. A more physical simulation is based on stochastic cellular automata: the cellular automata uses rules, such as Hale's and Joyce's law, for sunspot evolution, motion and conversion to energy (which leads to a flare and CME and thus to a SEP), which can be derived from sunspot observations. Stochastics enter in such a way that each solar cycle requires a stochastic seed of sunspot pairs and that the emergence of new sunspot pairs is determined stochastically from observed distributions (Poppenburg, 2006). Since the model 'destroys' sunspots by merging, it directly relates changes in sunspot number to energy released in flares and thus available for particle acceleration.

Although geomagnetic polarity transitions as such are well documented in paleomagnetic records, it is very challenging to reconstruct the evolution of the field coefficients during such a reversal. Leonhardt and Fabian (2007) presented a most promising and very complete approach to the problem of estimating the coefficients for the Matuyama-Brunhes reversal. The Leonhardt-Fabian model is based on high-quality paleomagnetic records and convincingly passed independent verification tests. Based on Leonhardt-Fabian coefficients large-scale magnetohydrodynamic simulations of the paleomagnetosphere during the Matuyama-Brunhes reversal for different solar wind conditions can be performed. Regions of open field lines on the Earth's surface could be identified by means of a three-dimensional field line tracing tool, and the resulting ionization in these regions and the effects on atmospheric chemistry and composition during the reconstructed Matuyama-Brunhes reversal can be reconstructed by means of the methods described in this chapter.

# References

Baumjohann, W. and Treumann, R. A. (1997). *Basic Space Plasma Physics*. Imperial Coll. Press, London.

Bethe, H. A. (1950). The range-energy relation for slow alpha-particles and protons in air. *Rev. Mod. Phys.*, 22:213.

Bieber, J. W., Eroshenko, E., Evenson, P., Flueckiger, E. O., and Kallenbach, R. (2000). *Cosmic rays and Earth*. Kluwer, Dordrecht.

Biernat, H. K., Kömle, N. I., and Lichtenberger, H. I. M. (1985). Analytical two-dimensional model of a quadrupole magnetosphere. *Planet. Space Sci.*, 33:45–52.

Bornebusch, J. P. (2005). *Asymmetrie in der Verteilung prezipierender Teilchen in den Polkappen*. Diploma thesis, Univ. of Osnabrück.

Bornebusch, J. P., Wissing, J. M., and Kallenrode, M.-B. (2008). Influences on polar particle precipitation. *Adv. Space Res.*, in press.

Callis, L. B., Natarajan, M., Lambeth, J. D., and Baker, D. N. (1998). Solar atmospheric coupling by electrons (solace), 2. Calculated stratospheric effects of precipitating electrons, 1979–1988. *J. Geophys. Res.*, 103:28 241.

Cane, H. V., McGuire, R. E., and von Rosenvinge, T. T. (1986). Two classes of solar energetic particle events associated with impulsive and long-duration soft x-ray flares. *Astrophys. J.*, 301:448.

Chipperfield, M. (1999). Multiannual simulations with a three-dimensional chemical transport model. *J. Geophys. Res.*, 104:1781–1805.

Clement, B. M. (1991). Geographical distribution of transitional VGP's: Evidence for non-zonal equatorial symmetry during the Matuyama-Brunhes geomagnetic reversal. *Earth Planet. Sci. Lett.*, 104:48–58.

Clement, B. M. and Kent, D. V. (1985). A comparison of two sequential polarity transitions (upper Olivai and lower Jaramillo) from the southern hemisphere. *Phys. Earth Planet. Inter.*, 39:310–313.

Cockell, C. (1999). Crises and extinction in the fossil record – a role for ultraviolet radiation? *Paleobiology*, 25:212–225.

Coe, R. S. and Prevot, M. (1989). Evidence suggesting extremely rapid field variation during during a geomagnetic reversal. *Earth Planet. Sci. Lett.*, 92:292–298.

Coe, R. S., Prevot, M., and Camps, P. (1995). New evidence for extraordinarily rapid change of the geomagnetic field during a reversal. *Nature*, 374:687–692.

Constable, C. and Parker, R. (1988). Statistics of the geomagnetic secular variation for the past 5 My. *J. Geophys. Res.*, 93:11569–11581.

Crutzen, P. and Bruehl, C. (1996). Mass extinctions and supernova explosions. *Proc. Natl. Acad. Sci. USA*, 93:1582–1584.

Crutzen, P. J., Isaksen, I. S. A., and Reid, G. C. (1975). Solar Proton Events: Stratospheric Sources of Nitric Oxide. *Science*, 189:457–459.

Ellis, J. and Schramm, D. (1995). Could a nearby supernova explosion have caused a mass extinction? *Proc. Natl. Acad. Sci. USA*, 92:235–238.

Fanselow, J. L. and Stone, E. C. (1972). Geomagnetic cutoffs for cosmic ray protons for seven energy intervals between 1.2 and 39 Mev. *J. Geophys. Res.*, 7:3999–4009.

Flueckiger, E. O. and Kobel, E. (1990). Aspects of combining models of the Earth's internal and external magnetic field. *J. Geomag. Geoelectr.*, 42:1123–1136.

Glassmeier, K., Vogt, J., Stadelmann, A., and Buchert, S. (2004). Concerning long-term geomagnetic variations and space climatology. *Ann. Geophys.*, 22:3669–3677.

Glassmeier, K.-H. (1997). The Hermean magnetosphere and its ionosphere-magnetosphere coupling. *Planet. Space Sci.*, 45:119–125.

Glatzmaier, G. A. and Roberts, P. H. (1995). A three-dimensional self-consistent computer simulation of a geomagnetic field reversal. *Nature*, 377:203–209.

Glatzmaier, G. A. and Roberts, P. H. (1996). Rotation and magnetism of Earth's inner core. *Science*, 274:1887–1891.

Gombosi, T. I., DeZeeuw, D. L., Groth, C. P. T., Powell, K. G., and Song, P. (1998). The length of the magnetotail for northward IMF: Results of 3D MHD simulations. In Chang, T. and Jasperse, J. R., editors, *Physics of Space Plasmas*, Vol. 15, pages 121–128. Mass. Inst. of Technol. Center for Theoretical Geo/Cosmo Plasma Physics, Cambridge, Mass.

Guyodo, Y. and Valet, J. P. (1999). Global changes in intensity of the earths magnetic field during the 800 kyr. *Nature*, 399:249–252.

Hauglustaine, D. and Gerard, J.-C. (1990). Possible composition and climatic changes due to past intense energetic particle precipitation. *Ann. Geophys.*, 8:87–96.

Hill, T. W., Dessler, A. J., and Wolf, R. A. (1976). Mercury and Mars – The role of ionospheric conductivity in the acceleration of magnetospheric particles. *Geophys. Res. Lett.*, 3:429–432.

Jackman, C., DeLand, M., Labow, G., Fleming, E., Weisenstein, D., Ko, M., Sinnhuber, M., Anderson, J., and Russell, J. (2005a). The influence of the several very large solar proton events in years 2000–2003 on the neutral middle atmosphere. *Adv. Space Res.*, 35:445–450.

Jackman, C., Douglass, A., Rood, R., McPeters, R., and Meade, P. (1990). Effect of solar proton events on the middle atmosphere during the past two solar cycles as computed using a two-dimensional model. *J. Geophys. Res.*, 95:7417–7428.

Jackman, C. H., DeLand, M. T., Labow, G. J., Fleming, E. L., Weisenstein, D. K., Ko, M. K. W., Sinnhuber, M., and Russell, J. M. (2005b). Neutral atmospheric influences of the solar proton events in October-November 2003. *J. Geophys. Res.*, 110:A09S27.

Jackman, C. H., Fleming, E. L., and Vitt, F. M. (2000). Influence of extremely large solar proton events in a changing stratosphere. *J. Geophys. Res.*, 105:11 659–11 670.

Jackman, C. H., McPeters, R. D., Labow, G. J., Praderas, C. J., and Fleming, E. L. (2001). Measurements and model predictions of the atmospheric effects due to the july 2000 solar proton event. *Geophys. Res. Lett.*, 28:2883.

Jackson, A. (1995). Storm in a lava flow? *Nature*, 377:685–686.

Kabin, K., Gombosi, T. I., DeZeeuw, D. L., and Powell, K. G. (2000). Interaction of mercury with the solar wind. *Icarus*, 143:397–406.

Kallenrode, M.-B. (1998). *Space Physics: An Introduction to Plasmas and Particles in the Heliosphere and Magnetospheres*. Springer-Verlag, Berlin.

Kallenrode, M.-B. (2003). Current views on impulsive and gradual solar energetic particle events. *J. Phys. G*, 29:965.

Kallenrode, M.-B., Cliver, E. W., and Wibberenz, G. (1992). Composition and azimuthal spread of solar energetic particles from impulsive and gradual flares. *Astro. Phys. J.*, 391:370.

Kinnersley, J. (1996). The climatology of the stratospheric thin air model. *Q.J.R. Meteorol. Soc.*, 122:219–252.

Krillke, C. (2006). *Teilcheneinfall in der Polkappe*. Diploma thesis, University of Osnabrück.

Lary, D. (1997). Catalytic destruction of stratospheric ozone. *J. Geophys. Res.*, 102:21515–21526.

Lemaire, J. F., Heyndrerickx, D., and (eds.), D. N. B. (1997). *Radiation belts – models and standards*. American Geophysical Union, Washington.

Leonhardt, R. and Fabian, K. (2007). Paleomagnetic reconstruction of the global geomagnetic field evolution during the matuyama/brunhes transition: Iterative bayesian inversion and independent verification. *Earth Planet. Sci. Lett.*, 253:172–195.

Leske, R. A., Mewaldt, R. A., Stone, E. C., and von Rosenvinge, T. T. (2001). Observations of geomagnetic cutoff variations during solar energetic particle events and implications for the radiation environment at the space station. *J. Geophys. Res.*, 106:30011–30022.

Leubner, M. P. and Zollner, K. (1985). The quadrupole magnetopause. *J. Geophys. Res.*, 90:8265–8268.

McCracken, K. G., Dreschhoff, G. A. M., Smart, D. F., and Shea, M. (2001a). Solar cosmic ray events for the period 1561–1994, (2) the gleissberg periodicity. *J. Geophys. Res.*, 106:21 599–21 609.

McCracken, K. G., Dreschhoff, G. A. M., Zeller, E. J., Smart, D. F., and Shea, M. A. (2001b). Solar cosmic ray events for the period 1561–1994, (1) identification in polar ice. *J. Geophys. Res.*, 106:21 585–21 598.

McElhinny, M. W. and Senanayake, W. E. (1982). Variations in the geomagnetic dipole. I – the past 50,000 years. *J. Geomagn. Geoelectr.*, 34:39–51.

McKibben, R. B. (1987). Galactic cosmic rays and anomalous components in the heliosphere. *Rev. Geophys.*, 25(3):711.

Melott, A., Thomas, B., Hogan, D., Ejzak, L., and Jackman, C. (2005). Climatic and biogeochemical effects of a galactic gamma ray burst. *Geophys. Res. Lett.*, 32:L14808.

Merrill, R. T. and McFadden, P. L. (1999). Geomagnetic polarity transitions. *Rev. Geophys.*, 37:201–226.

Pavlov, A.A., Pavlov, A.K., Mills, M.J., Ostryakov, V.M., Vasiljev, G.I., and Toon, O.B. (2005). Catastrophic ozone loss during passage of the solar system through an interstellar cloud. *Geophys. Res. Lett.*, 32:L01815.

Poppenburg, J. (2006). *Simulation of the solar cycle based on a probabilistic cellular automaton*. BA thesis, University of Osnabrück.

Porter, H. S., Jackman, C. H., and Green, A. E. S. (1976). Efficiencies for production of atomic nitrogen and oxygen by relativistic proton impact in air. *J. Chem. Phys.*, 65:154–167.

Prinn, R. and B. Fegley, J. (1987). Bolide impacts, acid rain, and biospheric traumas at the cretaceous-tertiary boundary. *Earth Planet. Sci. Lett.*, 83:1–15.

Quack, M., Kallenrode, M.-B., von König, M., Künzi, K., Burrows, J., Heber, B., and Wolff, E. (2001). Ground level events and consequences for stratospheric chemistry. In *Proceedings of ICRC 2001*. Copernicus Gesellschaft.

Randall, C. E., Siskind, D. E., and Bevilacqua, R. M. (2001). Stratospheric $NO_x$ enhancements in the southern hemisphere vortex in winter/spring of 2000. *Geophys. Res. Lett.*, 28:2385–2388.

Reames, D. V. (1999). Particle acceleration at the sun and in the heliosphere. *Space Sci. Rev.*, 90:413.

Reid, G., Isaksen, I., Holzer, T., and Crutzen, P. (1976). Influence of ancient solar-proton events on the evolution of life. *Nature*, 259:177–179.

Rishbeth, H. (1985). The quadrupole ionosphere. *Ann. Geophys.*, 3:293–298.

Roelof, E. C. and Sibeck, D. G. (1993). Magnetopause shape as a bivariate function of interplanetary magnetic field $B_z$ and solar wind dynamic pressure. *J. Geophys. Res.*, 98:21421.

Rohen, G., von Savigny, C., Sinnhuber, M., Llewellyn, E. J., Kaiser, J. W., Jackman, C. H., Kallenrode, M.-B., Schröter, J., Eichmann, K.-U., Bovensmann, H., and Burrows, J. P. (2005). Ozone depletion during the solar proton events of October/November 2003 as seen by SCIAMACHY. *J. Geophys. Res.*, 110:A09S39.

Russell, C. T. (2001). Solar wind and interplanetary magnetic field: a tutorial. In Song, P., Singer, H. J., and Siscoe, G. L., editors, *Space Weather Geophys. Monogr. Ser.*, Volume 125, pp. 71. AGU, Washington, D.C.

Saito, T., Sakurai, T., and Yumoto, K. (1978). The Earth's palaeomagnetosphere as the third type of planetary magnetosphere. *Planet. Space Sci.*, 26:413–422.

Sander et al., S. S. (2000). Chemical kinetics and photochemical data for use in stratospheric modeling, evaluation no 13. *JPL Publications*, 00-3.

Sander et al., S. S. (2006). Chemical kinetics and photochemical data for use in stratospheric modeling, evaluation no 15. *JPL Publications*, 06-2.

Schröter, J., Heber, B., Steinhilber, F., and Kallenrode, M. B. (2006). Energetic particles in the atmosphere: A Monte-carlo simulation. *Adv. Space Res.*, 37:1597–1601.

Schwenn, R. (1990). Large-scale structure of the interplanetary medium. In Schwenn, R. and Marsch, E., editors, *Physics of the inner heliosphere*, Volume 1, pp. 99. Springer, Berlin.

Shea, M. A., Smart, D. F., and McCracken, K. G. (1965). A study of vertical cutoff rigidities using sixth degree simulations of the geomagnetic field. *J. Geophys. Res.*, 70:4117.

Siebert, M. (1977). Auswirkungen der säkularen Änderung des erdmagnetischen Hauptfeldes auf Form und Lage der Magnetosphäre und die Stärke der erdmagnetischen Aktivität. *Abh. Braunschweig. Wiss. Ges.*, 37:281–319.

Sinnhuber, M., Burrows, J. P., Chipperfield, M. P., Jackman, C. H., Kallenrode, M.-B., Künzi, K. F., and Quack, M. (2003). A model study of the impact of magnetic field structure on atmospheric composition during solar proton events. *Geophys. Res. Lett.*, 30:1818–1821.

Siscoe, G. L. (1979). Towards a comparative theory of magnetospheres. In *Solar System Plasma Physics*, Volume II, pp. 319–402. North-Holland Publ. Comp.

Siscoe, G. L. and Chen, C.-K. (1975). The paleomagnetosphere. *J. Geophys. Res.*, 80:4675–4680.

Siscoe, G. L. and Crooker, N. J. (1976). Auroral zones in a quadrupole magnetosphere. *J. Geomagn. Geoelectr.*, 28:1–9.

Siscoe, G. L., Erickson, G. M., Sonnerup, B. U. Ö., Maynard, N. C., Schoendorf, J. A., Siebert, K. D., Weimer, D. R., White, W. W., and Wilson, G. R. (2002). Hill model of transpolar potential saturation: Comparisons with MHD simulations. *J. Geophys. Res.*, 107:1075.

Smart, D. F. and Shea, M. A. (1972). Daily variation of electron and proton geomagnetic cutoffs calculated for fort churchill, Canada. *J. Geophys. Res.*, 77:4595–4601.

Smart, D. F. and Shea, M. A. (2001). A comparison of the tsyganenko model predicted and measured geomagnetic cutoff latitudes. *Adv. Space Res.*, 28:1733–1738.

Smart, D. F. and Shea, M. A. (2005). A review of geomagnetic cutoff rigidities for earth-orbiting spacecraft. *Adv. Space Res.*, 36:2012–2020.

Smart, D. F., Shea, M. A., and Flückiger, E. O. (2000). Magnetospheric Models and Trajectory Computations. *Space Sci. Rev.*, 93:305–333.

Solomon, S., Reid, G. C., Rusch, D. W., and Thomas, R. J. (1983). Mesospheric ozone depletion during the solar proton event of July 13, 1982. II – Comparison between theory and measurements. *Geophys. Res. Lett.*, 10:257–260.

Solomon, S., Rusch, D. W., Gérard, J. C., Reid, G. C., and Crutzen, P. J. (1981). The effect of particle precipitation events on the neutral and ion chemistry of the middle atmosphere: II. Odd hydrogen. *Planet. Space Sci.*, 29:885–893.

Solomon, S. C. (2001). Auroral particle transport using monte-carlo and hybrid methods. *J. Geophys. Res.*, 106:107.

Stadelmann, A. (2004). *Globale Effekte einer Erdmagnetfeldumkehr: Magnetosphärenstruktur und kosmische Teilchen*. Dissertation, Technische Universität Braunschweig.

Starchenko, S. and Shcherbakov, V. (1991). Inverse magnetosphere. *Doktorlady Akademii Nauk SSSR*, 321. in Russian.

Steinhilber, F. (2005). *Simulation der solaren Aktivität auf Zeitskalen von Solarzyklen bis zu Jahrhunderten*. Diploma thesis, University of Osnabrück.

Stephenson, J. A. E. and Scourfield, M. W. J. (1992). Ozone depletion over the polar caps caused by solar protons. *Geophys. Res. Lett.*, 19:2425–2428.

Størmer, C. (1955). *The Polar Aurora*. Clarendon Press, Oxford.

Svensmark, H. (2000). Cosmic rays and earth's climate. *Space Sci. Rev.*, 93:175–185.

Svensmark, H. and Friis-Christensen, E. (1997). Variation of cosmic ray flux and global cloud coverage—a missing link in solar-climate relationship,. *J. Atm. Solar-Terr. Phys.*, 59:1225–1232.

Swider, W. and Keneshea, T. (1973). Decrease of ozone and atomic oxygen in the lower mesosphere during a pct event. *Planet. Space Sci.*, 21:1969–1973.

Toon, O. (1997). Environmental perturbations caused by the impacts of asteroids and comets. *Rev. Geophys.*, 35:41– 8.

Ultré-Guérard, P. and Achache, J. (1995). Core flow instabilities and geomagnetic storms during reversals: The Steens Mountain impulsive field variations revisited. *Earth Planet. Sci. Lett.*, 135:91–99.

Usoskin, I. G., Marsh, N., Kovaltsov, G. A., Mursula, K., and Gladysheva, O. G. (2004). Latitudinal dependence of low cloud amount on cosmic ray induced ionization. *Geophys. Res. Lett.*, 31:L16109.

Vitt, F. M. and Jackman, C. H. (1996). A comparison of sources of odd nitrogen production from 1974 through 1993 in the earth's middle atmosphere as calculated using a two-dimensional model. *J. Geophys. Res.*, 101:6729.

Vogt, J. and Glassmeier, K.-H. (2000). On the location of trapped particle populations in quadrupole magnetospheres. *J. Geophys. Res.*, 105:13,063–13,071.

Vogt, J. and Glassmeier, K.-H. (2001). Modelling the paleomagnetosphere: strategy and first results. *Adv. Space Res.*, 28:863–868.

Vogt, J., Zieger, B., Glassmeier, K.-H., Stadelmann, A., Kallenrode, M.-B., Sinnhuber, M., and Winkler, H. (2007). Energetic particles in the paleomagnetosphere: reduced dipole configurations and quadrupolar contributions. *J. Geophys. Res.*, 112:6216, doi:10.1029/2006JA012224.

Vogt, J., Zieger, B., Stadelmann, A., Glassmeier, K.-H., Gombosi, T. I., Hansen, K. C., and Ridley, A. J. (2004). MHD simulations of quadrupolar paleomagnetospheres. *J. Geophys. Res.*, 109:A12221.

Voigt, G.-H., Behannon, K. W., and Ness, N. F. (1987). Magnetic field and current structures in the magnetosphere of Uranus. *J. Geophys. Res.*, 92:15,337–15,346.

Voigt, G.-H. and Ness, N. F. (1990). The magnetosphere of Neptune: its response to daily rotation. *Geophys. Res. Lett.*, 17:1705–1708.

Walt, M. (1994). *Introduction to geomagnetically trapped radiation*. Cambridge University Press, Cambridge.

Wendler, J. (2004). External forcing of the geomagnetic field ? implications for the cosmic ray flux – climate variability. *J. Atm. Terr. Phys.*, 66:1195–1203.

Williams, I. and Fuller, M. (1981). Zonal harmonic models of reversal transition fields. *J. Geophys. Res.*, 86:11,657–11,665.

Willis, D. M., Holder, A. C., and Dais, C. J. (2000). Possible configurations of the magnetic field in the outer magnetosphere during geomagnetic polarity reversals. *Ann. Geophys.*, 18:11–27.

Winkler, H., Sinnhuber, M., Notholt, J., Kallenrode, M.-B., Steinhilber, F., Vogt, J., Zieger, B., Glassmeier, K.-H., and Stadelmann, A. (2008a). Modelling impacts of geomagnetic field variations on middle atmospheric responses to solar proton events on long time scales. *J. Geophys. Res.*, 13:2302, doi:10.1029/2007JD008574.

Wissing, J. M. (2005). *Räumliche und zeitliche Verteilung prezipierender magnetosphärischer Teilchen*. Diploma thesis, University of Osnabrück.

Wissing, J. M., Bornebusch, J. P., and Kallenrode, M.-B. (2008a). Variation of energetic particle precipitation with local magnetic time. *Adv. Space Res.*, 41:1274–1278.

Wissing, J. M., Sinnhuber, M., Winkler, H., and Kallenrode, M.-B. (2008b). Total inventory of precipitating particles and atmospheric consequences: October/November 2003 revisited. *Geophys. Res. Lett.*, submitted.

World Meteorological Organisation (2000). *Scientific Assessment of Ozone Depletion: 1999.* WMO.

World Meteorological Organisation (2003). *Scientific Assessment of Ozone Depletion: 2002.* WMO.

Zieger, B., Vogt, J., and Glassmeier, K.-H. (2006a). Scaling relations in the paleomagnetosphere derived from MHD simulations. *J. Geophys. Res.*, 111:A06203.

Zieger, B., Vogt, J., Glassmeier, K.-H., and Gombosi, T. I. (2004). Magnetohydrodynamic simulation of an equatorial dipolar paleomagnetosphere. *J. Geophys. Res.*, 109:A07205.

Zieger, B., Vogt, J., Ridley, A. J., and Glassmeier, K.-H. (2006b). A parametric study of magnetosphere-ionosphere coupling in the paleomagnetosphere. *Adv. Space Res.*, 38:1707–1712.

# Index